Junge, Lukhaup

Automotive Engineering
Kraftfahrzeugtechnik

English   Englisch
German   Deutsch

terminus

Hans-Dieter Junge, Dieter Lukhaup

Pocket Dictionary of
Automotive Engineering
English German

Taschenwörterbuch
Kraftfahrzeugtechnik
Englisch Deutsch

Verlag für Architektur
und technische Wissenschaften
Berlin

CIP-Kurztitelaufnahme der Deutschen Bibliothek

Junge, Hans-Dieter:
Pocket dictionary of automotive engineering: Engl.-German;
engl.-dt. = Taschenwörterbuch Kraftfahrzeugtechnik/Hans-Dieter Junge;
Dieter Lukhaup. –
Berlin: Verlag für Architektur u. Techn. Wiss. 1987.
   (Terminus pocket dictionary)
   ISBN 3-433-02815-X
   ISSN 0930-6846
NE: Lukhaup, Dieter:; HST

EDV-Verarbeitung und Lasersatz: U. Hellinger, D-6901 Heiligkreuzsteinach

Druck: Alphabet KG, D-1000 Berlin 36

Bindung: Reinhardt & Wasser, D-1000 Berlin 36

Printed in the Federal Republic of Germany

Vorwort

Die moderne Technik fordert eine immer engere Spezialisierung. Der Student ist schon während seines Studiums gezwungen, sich in fremdsprachige und speziell in englische Spezialliteratur einzulesen. Er bemerkt bald, daß zwischen seinem Schulwissen und der Fachsprache ein weiter Unterschied besteht. Viele Vokabeln erhalten plötzlich eine neue Bedeutung, andere kommen hinzu.

Dieses Taschenwörterbuch soll dem Studenten und dem Berufsanfänger den Übergang zur Fachsprache seines Spezialgebietes erleichtern. Es soll aber auch dem Übersetzer bei der Übersetzung von Spezialliteratur helfen.

Das Taschenwörterbuch enthält neben einem ausgewählten Querschnitt der wichtigsten Fachbegriffe der Kraftfahrzeugtechnik, teilweise mit Definitionen oder Erklärungen versehen, zahlreiche Beispiele für die Bildungsgesetze von Wortkombinationen. Außerdem wurde eine Anzahl allgemeinsprachlicher Begriffe aufgenommen, für die erfahrungsgemäß gerade vom Fachmann oft nach einem für den technischen Text geeigneten Äquivalent gesucht wird.

Das Taschenwörterbuch kann ein großes Fachwörterbuch nicht ersetzen, in vielen Fällen jedoch wird es dieses ergänzen.

Autoren und Verlag wünschen, daß dieses Buch seinen Zweck erfüllen möge, nämlich jedem Benutzer bei minimalem Aufwand eine maximale Hilfe zu sein.

Herrn Dipl.-Übers. Bernd Schorr, München, sei für zahlreiche wertvolle Hinweise besonders gedankt.

Benutzungshinweise

Im terminus-Taschenwörterbuch sind die Eintragungen in alphabetisch geordneten Wortnestern zusammengestellt, wobei darauf Wert gelegt ist, möglichst unterschiedliche Kombinationen des Nestwortes aufzunehmen, um die Bildungsgesetze der Wortkombinationen sichtbar werden zu lassen.

Innerhalb der Wortstellen sind die Eintragungen in der Quellsprache hervorgehoben. Eintragungen in runden Klammern können das vorangehende Wort ersetzen. Einträge in eckigen Klammern entfallen oft, insbesondere im amerikanischen Englisch. Angaben in *Kursivschrift* helfen bei der Zuordnung des Begriffes zu einem speziellen Anwendungsfall oder zu einem bestimmten technischen Bereich.

Es sei noch darauf hingewiesen, daß es im amerikanischen Englisch zahlreiche Abweichungen gegenüber dem „klassischen" Englisch gibt. Wie in der technisch-wissenschaftlichen Literatur üblich, wird im Zweifelsfall der amerikanischen Schreibweise der Vorzug gegeben, im „klassischen" Englisch verwendete abweichende Schreibweisen wurden, soweit sie weit verbreitete Anwendung finden, angemerkt.

Im Alphabet sind die Buchstaben ä, ö und ü wie a, o und u eingeordnet. ß ist wie ss behandelt.

Introduction

Advanced technology requires more and more specialized technicians, students, engieneers and scientists. The student must unavoidably read foreign-language technical and scientific literature. Soon he must recognize a wide gap between his acquired knowledge and the language of technology and science. Many words change their meaning, others are completely new.

This Pocket Dicitonary is thought to facilitate the jump from the every-day language to the scientific-technical language. Furthermore, it should be a helpful tool for the translator who deals with specialized literature.

This Pocket Dictionary summarizes the most important and frequently used terms in the field of automotive engineering, partially supplemented by definitions or other statements. Numerous examples of word combinations are included for better understanding of the linguistic laws. Some general terms are also included which have a special meaning in the scope of this book.

This Pocket Dictionary of Automotive Engineering is not intended to replace a large specialist dictionary. Nevertheless, it may supplement such a book. A further volume contains the German-English translations.

Publishers and authors hope the Pocket Dictionary may fulfil its purpose, namely, to be an effective tool with minimum expense for every user.

Many thanks to Dipl.-Übersetzer Bernd Schorr, Munich, for numerous valuable comments.

Instructions for use

All entries are alphabetically ordered and, if appropriate, combined with groups of related words. Groups of word combinations are thought to be useful for the understanding of linguistic laws.

Within the entries the source-language terms are bold-faced. Terms within parantheses can be used to replace the term before. Terms in square brackets are dropped sometimes, mostly in American English. Annotations in italics may be helpful in relating the entry to a special technical scope.

In most cases the American version of spelling is preferred as American English is used more and more in science and technology.

In the German alphabet the mutated vowels ä, ö and ü are ordered as a, o and u, ß is ordered as ss.

A-post Vorderwandsäule f, Windschutzscheibensäule f
ability Fähigkeit f
  starting ability Startfähigkeit f, Startvermögen n
abrasion Abrieb m, Verschleiß m
abrasive resistance Abriebfestigkeit f
absorb/to absorbieren, aufsaugen
absorber Absorbierer m, Dämpfer m
  shock absorber Stoßdämpfer m
  shock absorber inclination Stoßdämpferneigung f
  shock absorber mounting Stoßdämpferlager n
  shock absorber plunger Stoßdämpferkolben m
  telescopic shock absorber Teleskopstoßdämpfer m
  vibration absorbing mo[u]ld Schwingungsdämpfungsglied n
accelerate/to beschleunigen
accelerating pump Beschleunigungspumpe f
acceleration Beschleunigung f
accelerator Gaspedal n
  accelerator pedal Gaspedal n
  accelerator pump Beschleunigungspumpe f
  ignition accelerator Zündbeschleuniger m
accessories Zubehör n
accidental ground [unerwünschter] Masseschluß m
accumulator Akkumulator m, Batterie f
  accumulator cell Sammlerzelle f
  accumulator switch Batterieumschalter m
  spring energy accumulator Federspeicher m
acid Säure f
  acid density Säuredichte f
  acid-proof säurebeständig, säurefest
  acid-proof grease Säureschutzfett n
  acid-resisting säurebeständig
  acid tester Säureprüfer m
action Tätigkeit f, Handeln n, Funktion f, Funktionieren n, Betrieb m
  castor action Nachlauf m
  radius of action Fahrbereich m, Reichweite f, Aktionsradius m
  scavenging action Spülwirkung f
activated charcoal filter Aktivkohlefilter n
actuate/to betätigen, erregen *(z.B. Relais)*
actuating Betätigungs-, Schalt-

actuating bolt Schaltstift m
actuating cam Antriebsnocken m, Betätigungsnocken m
adapt/to anpassen
adaptability Anpassungsfähigkeit f
adapter Paßstück n, Zwischenstück n, Adapter m
additional zusätzlich
  additional air Zusatzluft f, Sekundärluft f
  additional lever Zusatzhebel
adhesion Adhäsion f, Bodenhaftung f
  adhesion power Haftfestigkeit f
adjust/to einstellen, verstellen
adjustable einstellbar
  adjustable anti-roll bar einstellbarer Stabilisator m
  adjustable resistor Stellwiderstand m
  adjustable seat verstellbarer Sitz m
  adjustable spot lamp Suchscheinwerfer m, Sucher m
adjuster Regler m, Steller m, Verstellelement n
  leaf adjuster Blattrichter m
  seat adjuster Sitzversteller m
  slack adjuster Leerwegeinsteller m, Gestängesteller m *(Bremse)*
adjusting Einstellen n, Stell-
  adjusting angle Einstellwinkel m
  adjusting body Verstellkörper m
  adjusting bridge Einstellbrücke f
  adjusting bush Einstellbüchse f
  adjusting device Einstellvorrichtung f
  adjusting disk Verstellscheibe f, Einstellscheibe f
  adjusting flange Einstellflansch m
  adjusting grip Stellgriff m
  adjusting lever Einstellhebel m, Verstellhebel m, Regulierhebel m
  adjusting nut Stellmutter f, Einstellmutter f
  adjusting plate Einstellblech n, Einstellplatte f
  adjusting ring Stellring m
  adjusting screw Einstellschraube f, Stellschraube f,
  Regulierschraube f
  adjusting sleeve Verstellmuffe f, Einstellhülse f, Stellhülse f,
  Nachstellmuffe f
  coarse adjusting Grobeinstellung f
  fine adjusting Feineinstellung f

fine adjusting nut Feineinstellmutter f
adjustment Verstellung f, Justierung f, Einstellung f, Regulierung f, Nachstellen n
adjustment angle Einstellwinkel m
adjustment mark Einstellmarke f
adjustment screw Nachstellschraube f, Einstellschraube f, Stellschraube f
adjustment spring Angleichfeder f, Stellfeder f, Paßfeder f
angular adjustment Winkeleinstellung f
contact breaker adjustment Unterbrechereinstellung f
height adjustment Höhenjustierung f
lateral adjustment Seitenjustierung f
range of adjustment Verstellbereich m, Einstellbereich m
rich adjustment fette Gemischeinregulierung f
seat height adjustment Sitzhöhenverstellung f
spark adjustment Zündverstellung f, Zündzeitpunktanpassung f
admission Einlaß m, Zuführung f, Zuleitung f
mixture admission Gemischzuführung f
advance Vorlauf m, Vorschub m, Voreilung f
advance angle Verstellwinkel m
advance characteristic Verstellinie f, Verstellkennlinie f
advance device (mechanism) Zündversteller m
centrifugal advance mechanism Fliehkraft-Zündversteller m
centrifugal spark advance Fliehkraftverstellung f
full advance volle Frühzündung f
ignition advance Zündzeitpunktvorgabe f
injection advance Einspritzvoreilung f, Voreinspritzung f
injection advance device Einspritzversteller m
spark advance Zündzeitpunktvorgabe f
vacuum advance Unterdruckverstellung f
vacuum advance mechanism Unterdruckzündversteller m
vacuum advance unit Frühdose f
vacuum ignition advance Unterdruckzündverstellung f
vacuum spark advance Unterdruckzündverstellung f
advanced ignition Vorzündung f, Frühzündung f
aerial Antenne f *(s.a. antenna)*
fender aerial Kotflügelantenne f
under-car aerial Chassisantenne f
aerodynamic aerodynamisch

aerodynamic drag Luftwiderstand m
aerodynamic energy Strömungsenergie f
aerodynamic force Luftwiderstandkraft f
coefficient of aerodynamic drag Luftwiderstandsbeiwert m
afterburning Nachverbrennung f
aftertreatment Nachbehandlung f *(des Abgases)*
agent Agens n, Mittel n
  cleaning agent Reinigungsmittel n
  washing agent Waschmittel n
aging Alterung f
aggregate Aggregat n, Gruppe f
  ventilator aggregate Lüfteraggregat n
ahead vorwärts, vorn
  ahead speed Vorwärtsgang m
  proceed ahead/to geradeaus weiterfahren
  run straight ahead/to geradeausfahren
aid Hilfe f
  cold-starting aid Kaltstarthilfe f
  first aid cushion Erste-Hilfe-Kissen n
  first aid kit Erste-Hilfe-Ausrüstung f
  ignition aid Zündhilfsmittel n, Starthilfe f
  starting aid Starthilfe f
air Luft f
  air baffle Luftleitblech n
  air bleed opening Nebenlufteinlaß m
  air bleed valve Belüftungs-[Entlüftungs-]ventil n
  air chamber Luftbehälter m, Luftkammer f
  air charge tube Luftladerohr n
  air cleaner bracket Luftansaugfilterstutzen m
  air cock Lufthahn m
  air conditioner Klimaanlage f
  air compressor Luftkompressor m
  air-conditioning Klimatisierung f
  air-conditioning compressor Kältekompressor m
  air-conditioning plant Klimaanlage f
  air contamination Luftverschmutzung f
  air correction jet Ausgleichluftdüse f, Luftkorrekturdüse f
  air deflector Leitblech n
  air density Luftdichte f

air excess   Luftüberschuß m
air excess ratio   Luftüberschußzahl f
air filter   Luftfilter n,m
air filter element   Luftfiltereinsatz m
air filter intake silencer   Luftfilteransauggeräuschdämpfer m
air flap   Luftklappe f
air flow rate   Luftdurchsatz m
air flow sensor   Luftmengenmesser m, Luftmesser m
air flow sensor plate   Stauscheibe f
air fuel mixture   Kraftstoff-Luft-Gemisch n
air-fuel ratio   Mischungsverhältnis n
air gap   Luftspalt m
air guide sheet   Luftleitblech n
air gulp valve   Abmagerungsventil n *(gegen Auspuff knallen)*
air hole   Entlüftungsloch n
air-inflation indicator   Füllluftmesser m
air injection   Lufteinblasung f, Sekundärlufteinspeisung f
air injection pump   Lufteinblaspumpe f,
Sekundärlufteinblaspumpe f
air injection rate   Lufteinblasmenge f, Sekundärluftmenge f
air injection tube   Lufteinblasrohr n, Sekundärlufteinblasrohr n,
Luftzugaberohr n
air inlet   Lufteinlaß m
air inlet tube   Lufteintrittstutzen m, Luftansaugrohr n
air intake   Lufteinlaß m
air jet   Luftdüse f
air manifold   Luftverteilerrohr n
air mass   Luftmenge f
air-operated cable brake   Seildruckluftbremse f
air-operated linkage brake   Gestängedruckluftbremse f
air plenum   Luftkammer f *(beim Katalysator)*
air pocket   Lufteinschluß m, Lufttasche f
air pollution   Luftverschmutzung f
air pressure cylinder   Luftdruckzylinder m
air pressure ga[u]ge   Druckluftmanometer n; Luftdruckprüfer m
(Reifendruck)
air-pressure assisted steering   Lenkung f mit pneumatischer
Lenkhilfe, pneumatische Servolenkung f
air pump   Luftpumpe f

air relief line   Entlüfterleitung f
air relief pipe   Entlüfterrohr n
air shutter plate   Starterklappe f, Luftklappe f
air slide   Luftschieber m
air strangler   Lufteinlaßklappe f, Starterdrosselklappe f
air-supply line   Fülleitung f
air swirl   Luftdrall m, Luftturbulenz f, Luftverwirbelung f
air temperature sensor   Lufttemperatursensor m
air-tight   luftdicht
air valve   Luftventil n
air vent   Entgasungsöffnung f, Luftloch n
air vent connection   Entlüfterstutzen m
air vent hose   Entlüfterschlauch m
air vent screw   Entlüfterschraube f, Entlüftungsschraube f
(Batterie)
ambient air   Umgebungsluft f
auxiliary air reservoir   Zusatzluftbehälter m
auxiliary air valve   Hilfsluftventil n
compressed air   Druckluft f
compressed air bar   Druckluftleiste f
compressed air brake   Druckluftbremse f, pneumatische Bremse f
compressed air cleaner   Druckluftfilter n,m
compressed air container   Druckluftbehälter m
compressed air cylinder   Druckluftzylinder m
compressed air hose   Druckluftschlauch m
compressed air line (pipe)   Druckluftleitung f
cooling-air stream   Kühlluftstrom m
excess air   Luftüberschuß m
exhaust-operated air heating   Abgasheizung f
fresh air screen   Frischluftgitter n
fuel-air ratio control unit   Gemischregler m
hot air   Warmluft f
hot air blower   Warmluftgebläse n
hot air flap   Warmluftklappe f
hot air funnel   Warmlufttrichter m
hot air hose   Warmluftschlauch m
hot air vent   Warmluftdüse f
induction air-swirl   Wirbelung f durch Ansaugluft
inlet air   Ansaugluft f

intake air   Ansaugluft f
intake air heater   Ansaugluftvorwärmer m,
Vergaserluftvorwärmer m
leak air   Falschluft f
main air reservoir   Hauptluftbehälter m *(Bremse)*
oil-wetted air cleaner   Naßluftfilter m,n
preliminary air tank   Vorluftbehälter m
scavenging air   Spülluft f
starting-air valve   Anlaßventil n
oil bath air filter   Ölbadluftfilter m,n
spark plug air gap   Elektrodenabstand m
align/to   ausfluchten, ausrichten, vermessen, einstellen *(bei
Radaufhängung)*
aligning   ausrichtend
self-aligning   selbstausrichtend
self-aligning ball bearing   Pendelkugellager n
self-aligning ring   Pendellagerring m
self-aligning roller bearing   Pendelrollenlager n
self-aligning torque   Rückstellmoment n *(Lenkung)*
alignment   Ausrichten n
track alignment   Radausfluchtung f
wheel alignment   Radeinstellung f, Spureinstellung f
all   ingesamt, voll-
all-independent suspension   Rundum-Einzelradaufhängung f
all-loss lubrication   Durchlaufschmierung f, verlorene Schmierung f
all-purpose tire   Allzweckreifen m
all-round glazing   Rundumverglasung f
all-season tire   Ganzjahresreifen m
all-steel body   Ganzstahlkarosserie f
all-synchromesh transmission   vollsynchronisiertes Getriebe n
all-wheel drive   Allradantrieb m
alloy   Legierung f
light alloy wheel   Leichtmetallrad n
alternating   abwechselnd, Wechsel-
alternating-current dynamo   Wechselstromdynamo m
alternating-current generator   Wechselstromlichtmaschine f
alternation of polarity   Polwechsel m
alternator   Wechselstromgenerator m, Wechselstromlichtmaschine f
three-phase alternator   Drehstromgenerator m

alumin[i]um Aluminium n
  alumin[i]um alloy Aluminiumlegierung f
  alumin[i]um casting Aluminiumguß m
  alumin[i]um coating Aluminiumüberzug m
  alumin[i]um disk Aluminiumscheibe f
  alumin[i]um housing Aluminiumgehäuse n
  alumin[i]um vapo[u]r plating Aluminiumbedampfung f
  alumin[i]um wheel Aluminiumrad n, Alurad n
amber light Gelblicht n
ambient air Umgebungsluft f
amperage Stromstärke f, Amperezahl f
ampere-hour Amperestunde f
amplification Verstärkung f
amplifier Verstärker m
amplify/to verstärken
anchor Anker m
angle Winkel m
  adjustment angle Einstellwinkel m
  advance angle Verstellwinkel m
  angle between cranks Kurbelversetzung f
  angle bolt Winkelbolzen m
  angle bracket Stützwinkel m, Befestigungswinkel m, Verstärkungswinkel m
  angle gear Winkelgetriebe n
  angle iron Winkeleisen n
  angle joint rod Winkelgelenkstange f
  angle of approach vorderer Überhangwinkel m, vorderer Böschungswinkel m, Anstellwinkel m *(aerodynamisch)*
  angle of connecting rod Pleuelstangenausschlag m
  angle of delay Zündverzugswinkel m
  angle of departure hinterer Überhang m
  angle of incline Neigungswinkel m, Überhangwinkel m, hinterer Böschungswinkel m
  angle of lock Ausschlagwinkel m *(Lenkung)*, Einschlagwinkel m *(Räder)*
  angle pinion Winkelrad n
  angle plate Winkelplatte f
  angle ratchet wheel Winkel-Sperrad n
  angle wrench Winkelschlüssel m

approach angle vorderer Überhangwinkel m, vorderer Böschungswinkel m, Anstellwinkel m *(aerodynamisch)*
bearing angle Lagerwinkel m
camber angle Radsturzwinkel m
castor angle Nachlaufwinkel m
closing angle Abschlußwinkel m
connecting angle Anschlußwinkel m, Verbindungswinkel m
contact angle Kontaktwinkel m
fixing angle Befestigungswinkel m
guide angle Führungswinkel m
holding angle Haltewinkel m
king-pin angle Radspreizungswinkel m
lock angle Einschlagwinkel m *(Räder)*
mounting angle Montagewinkel m
pressure angle Eingriffswinkel m
reinforcing angle Verstärkungswinkel m
return guide angle Rückführungswinkel m
setting angle Einstellwinkel m
slip angle Schrägaufwinkel m
stiffening angle Versteifungswinkel m
stop angle Anschlagwinkel m
supporting angle Auflagewinkel m, Haltewinkel m
thrust angle Stützwinkel m
tightening angle Spannwinkel m
timing angle Verstellwinkel m *(Zündung)*
trail angle Spurwinkel m
wide-angle lighting fitting Breitstrahler m
wiping angle Wischwinkel m
angular Winkel-
  angular adjustment Winkeleinstellung f
  angular bracket Winkelstütze f
  angular connection Winkelanschluß m
  angular deceleration Winkelverzögerung f
  angular displacement Beugungswinkel m *(Kurvenfahrt)*
  angular drive Winkeltrieb m
  angular socket Winkelsteckdose f
annular Ring-
  annular lamp holder Ringfassung f
  annular valve Ringventil n

antenna Antenne f *(s.a. aerial)*
  bar antenna Stabantenne f
  loop antenna Schleifenantenne f
anti entgegengerichtet, Anti-
  anti-block system Antiblockiersystem n
  anti-clockwise rotation Linksdrehung f, Linkslauf m
  anti-dazzle blendfrei
  anti-dazzle device Blendschutz m
  anti-dazzle mirror Abblendspiegel m *(elektrisch)*
  anti-dazzle switch Motorradabblendschalter m
  anti-drive control Bremswirkausgleich m
  anti-drive end Stirnseite f, Motorstirnseite f
  anti-freeze Frostschutz m
  anti-frost container Frostschutzbehälter m
  anti-frost screen Frostschutzscheibe f
  anti-glare device Blendschutz m
  anti-interference cable Entstörkabel n
  anti-interference capacitor Entstörkondensator m
  anti-tampering plug Sicherungskappe f *(am Einspritzsystem)*
antiknock Antiklopf-
  antiknock additive Antiklopfzusatz m, Klopfbremse f *(bei Kraftstoff)*
  antiknock agent Antiklopfzusatz m, Klopfbremse f *(bei Kraftstoff)*
  antiknock characteristics Klopffestigkeit f *(bei Kraftstoff)*
  antiknock property Klopfverhalten n
  antiknock quality Klopffestigkeit f
antipole Gegenpol m
antiroll bar Querstabilisator m
antiskid Gleitschutz-
  antiskid chain Gleitschutzkette f
  antiskid tire Haftreifen m, rutschsicherer Reifen m
aperture Öffnung f
apex Kolbenecke f, Rotorecke f *(beim Kreiskolbenmotor)*
  apex seal Dichtleiste f *(beim Kreiskolbenmotor)*
apparatus Apparat m, Mechanismus m
  switch apparatus Schaltapparat m
applied ignition Fremdzündung f
apply/to anlegen *(von außerhalb)*

approach mark Zulassungszeichen n
aquaplaning Aquaplaning n *(Verlust der Lenkfähigkeit durch Gleiten auf Wasserfilm)*
arc Bogen m, Lichtbogen m
  arc lever Bogenhebel m
  arc welding Lichtbogenschweißen n, Elektroschweißen n
  pole arc Polbogen m
  pressure arc Andrückbogen m
arcing Funkenbildung f, Funkenüberschlag m
area Fläche f, Bereich m
  exit area of light Lichtaustrittfläche f
  wiped area Wischfeld n
arm Arm m, Ausleger m
  arm support Armstütze f
  bearing arm Lagerarm m
  contact arm Unterbrecherhebel m
  crank arm Kurbelarm m
  distributor (rotor) arm Verteilerfinger m
  drop arm Lenkstockhebel m, Lenkspurhebel m
  fixing arm Befestigungsarm m
  idler arm Lenkzwischenhebel m
  indicator arm Blinkerarm m
  lever arm Hebelarm m
  longitudinal control arm Längslenker m
  pedal arm Fußhebelarm m
  pitman arm Lenkstockhebel m, Lenkschneckenarm m
  rotor arm Verteilerfinger m
  screen wiper arm Scheibenwischerarm m
  shifting arm Schaltarm m
  supporting arm Stützarm m, Tragarm m
  track [control] arm Spurstangenhebel m
  trafficator arm Blinkerarm m
  trailing arm Trapezlenker m
  wiper arm Wischerarm m
armature Anker m, Rotor m
  armature brake Ankerbremse f
  armature coil Ankerspule f, Ankerwicklung f
  armature conductor Ankerleiter m
  armature shaft Ankerwelle f

armature short-circuiting brake   Ankerkurzschlußbremse f
armature spider   Ankerkörper m
armature spindle   Ankerwelle f
alternator armature   Wechselstromanker m
ignition armature   Zündanker m
sliding armature starter   Schubankeranlasser m
starter armature   Anlasseranker m
armrest   Armlehne f, Armstütze f
arrangement   Anordnung f
  cylinder arrangement   Zylinderanordnung f
  four-wheel steer arrangement   Vierradsteuerung f
  pole arrangement   Polanordnung f
arrest/to   arretieren, zurückhalten
arrester   Ableiter m; Sperreinrichtung f
  spark arrester   Funkenlöscher m
articulated   Gelenk-
  articulated bus   Gelenkbus m, Gliederbus m
  articulated tube   Gliederschlauch m
  articulated vehicle   Gelenkfahrzeug n, Sattelschlepper m, Sattelzug m
articulation   Gelenk n, Gelenkverbindung f
  rubber articulation   Gummigelenk n
ash content   Aschegehalt m
ashtray   Aschenbecher m
aspect ratio   Breiten-Verhältnis n, Höhe-zu-Breite-Verhältnis n *(bei Reifen)*
aspirate/to   ansaugen *(bei Motor)*
assemble/to   montieren
assembly   Montage f, Zusammenbau m, Aggregat n, Anlage f, Baugruppe f
  assembly drawing   Montagezeichnung f
  assembly instructions   Montageanleitung f, Montageanweisung f
  assembly jig   Montagegestell n, Montagevorrichtung f
  assembly line   Montageband n, Fließband n
  drive assembly   Getriebe n
  pinion drive assembly   Ritzelbetrieb m
  spring assembly   Federpaket n, Federsatz m
assisted   unterstützt
  air pressure assisted steering   Lenkung f mit pneumatischer

Lenkhilfe, pneumatische Servolenkung f
hydraulically assisted steering Lenkung f mit hydraulischer
Lenkhilfe, hydraulische Servolenkung f
power-assisted steering Servolenkung f
servo-power assisted servounterstützt
asymmetric asymmetrisch
  asymmetric dimming light asymmetrisches Abblendlicht n
  asymmetric tire abrasion einseitiger Reifenabrieb m
  asymmetric tire wear einseitiger Reifenverschleiß m
asymmetry Asymmetrie f
atomization Zerstäubung f, Vergasung f
attachment Befestigung f
  attachment face Radanlagefläche f
  attachment plug Zwischenstecker m
  point of attachment Anschlußpunkt m
  three-point attachment Dreipunktbefestigung f
attraction Anziehung f
audible hörbar, akustisch
  audible signal[l]ing device akustischer Signalgeber m
  audible warning control Hupenbetätigung f
  audible warning device akustischer Signalgeber m
  audible wear indicator akustischer Verschleißanzeiger m
  electric-audible signal horn elektroakustisches Signalhorn n
audio frequency Tonfrequenz f
auto lock Autoschloß n
automatic automatisch
  automatic adjustment Selbstnachstellung f
  automatic advance automatische Zündverstellung f
  automatic advance mechanism Zündversteller m
  automatic antenna automatische Antenne f
  automatic choke Startautomatik f
  automatic circuit-breaker Selbstausschalter m
  automatic dimming automatisches Abblenden n
  automatic disconnection Selbstauslösung f
  automatic gearbox automatisches Getriebe n
  automatic headlamp aimer Leuchtweitenregulierung f
  automatic headlamp level[l]ing Leuchtweitenregulierung f
  automatic injection advance mechanism automatischer
Spritzversteller m *(bei Einspritzung)*

automatic machine Automat m
automatic release Selbstauslösung f
automatic seat back release automatische Rückenlehnenentriegelung f
automatic seat belt Automatikgurt m, automatisch anlegender Sitzgurt m
automatic seat belt adjustment automatische Sitzgurteinstellung f
automatic spark advance automatische Zündverstellung f
automatic switch automatischer Schalter m
automatic timing automatische Zündverstellung f
automatic transmission gear Automatikgetriebe n
automatic valve clearance adjustment automatische Ventilspieleinstellung f
automatic version Automatikausführung f
automatic warm-up [control] device Warmlaufautomatik f
fully automatic vollautomatisch
automotive Kraftfahrzeug-
automotive aftermarket Fahrzeugzubehörmarkt m
automotive body Karosserie f
automotive engineering Kraftfahrzeugtechnik f
automotive industry Automobilindustrie f, Kraftfahrzeugindustrie f, Automobilbau m
auxiliaries Nebenaggregate npl
engine auxiliaries Motorausrüstung f
auxiliary Hilfs-, Zusatz-
auxiliary air Zusatzluft f
auxiliary air bleed Zusatzluftdüse f
auxiliary air device Zusatzluftschieber m
auxiliary air fan Zusatzlüfter m
auxiliary air jet Zusatzluftdüse f
auxiliary air reservoir Zusatzluftbehälter m
auxiliary air valve Zusatzluftventil n
auxiliary fuel tank Kraftstoffzusatztank m, Kraftstoff-Reservebehälter m
auxiliary gearbox Verteilergetriebe n
auxiliary headlamp Zusatzscheinwerfer m
auxiliary idler shaft Hilfswelle f, zusätzliche Leerlaufwelle f *(Getriebe)*

auxiliary reduction gear   Vorgelege n
auxiliary shaft   Hilfswelle f, Vorgelegewelle f, Vorgelege n
auxiliary spark gap   Hilfsfunkenstrecke f
auxiliary spring   Hilfsfeder f
auxiliary switch   Hilfsschalter m
auxiliary transmission   Zusatzgetriebe n
auxiliary valve   Zusatzventil n
auxiliary winding   Hilfswicklung f
average   Mittelwert m, Mittel n
  average consumption   mittlerer Verbrauch m
  average output   mittlere Leistung f
  average speed   mittlere Geschwindigkeit f
axial   axial; Axial-
  axial ball bearing   Axialkugellager n
  axial bearing   Axiallager n
  axial bearing ring   Axiallagerring m
  axial joint   Axialdichtung f
  axial load   Axialbelastung f
  axial play   Axialspiel n
  axial pump   Axialpumpe f
  axial roller bearing   Axialrollenlager n
axis   Achse f
  axis of refraction   Brechungsachse f
  axis of revolution   Rotationsachse f
  focal axis   Brennachse f
  optical axis   optische Achse f
  rotational axis   Rotationsachse f
axle   Achse f
  axle base   Achsabstand m
  axle bearing   Achslager n
  axle bevel   Achskegelrad n
  axle drive   Achsantrieb m
  axle guide   Achslenker m
  axle half   Achshälfte f
  axle housing   Achsgehäuse n
  axle journal   Lagerhals m
  axle load   Achslast f
  axle load distribution   Achslastverteilung f
  axle pivot   Achszapfen m

axle shaft   Achswelle f
axle strut   Achsstrebe f
axle suspension   Achsaufhängung f
axle tube   Achsrohr n
axle weight   Achslast f
banjo axle   Banjoachse f
beam axle   Starrachse f
differential axle-drive bevel gear   Differentialantriebskegelrad n
driving axle   Antriebsachse f
flared tube of rear axle   Hinterachstrichter m
floating axle   Schwingachse f, fliegende Achse f
foot lever axle   Fußhebelwelle f
front axle   Vorderachse f
full-floating axle   Pendelachse f, vollfliegende Achse f, Steckachse f
guide axle   Führungsachse f
half axle   Halbachse f
inner axle shaft   Differentialseitenwelle f
outer axle shaft   Radantriebswelle f
rear axle   Hinterachse f
rear axle housing   Hinterachsgehäuse n
rigid beam axle   Starrachse f
ring of axle   Achsenring m
rocker lever axle   Kipphebelachse f, Schwinghebelachse f
shift axle   Schaltachse f
stub axle   Vorderachsschenkel m
supporting axle   Halterachse f
swing axle   Schwingachse f
trailing axle   Hinterachse f
back   Rück-
  back clearance   Radialspiel n *(beim Kolbenring)*
  back-coupling   Rückkopplung f
  back-light blinder   Rücklichtabdeckblende f
  back motion roller   Rücklaufrolle f
  back panel   Heck n *(Karosserie)*, Rückwand f
  back plate   Grundplatte f *(bei Trommelbremse)*
  back pressure   Staudruck m, Gegendruck m *(Abgas)*
  back pressure spring   Rückdruckfeder f
  back rest   Rückenlehne f
  set-back lever   Rückstellhebel m

backfire/to frühzünden, fehlzünden
backfire Frühzündung f, Fehlzündung f
backfit/to nachträglich ausrüsten
backing lamp Rückfahrscheinwerfer m
backrest Rückenlehne f
  backrest adjustment Rückenlehnenverstellung f
  backrest retainer Rückenlehnenhalterung f
badge Plakette f
baffle Leitblech n, Ablenkblech n
  baffle plate Prallblech n, Ablenkblech n, Leitblech n
  baffle ring Spritzring m
  air baffle Luftleitblech n
bag Luftsack m, Luftkissen n
  air bag system Luftsacksystem n
  tool bag Werkzeugtasche f
  wheel chock bag Radkeiltasche f
bajonet lock Bajonettverschluß m
baked enamel Einbrennlack m
bakelite Bakelit n
balance Gleichgewicht n, Waage f
  balance hole Auswucht f, Ausgleichbohrung f
  balance shaft Balancewelle f
  balance spring Balancefeder f
  spring balance Federwaage f, Federwippe f
balancing Ausgleich m
  balancing of masses Massenausgleich m
  balancing weight Ausgleichgewicht n, Auswuchtgewicht n
ball Kugel f, Kupplungskugel f *(bei Anhängevorrichtung)*
  ball-and-nut power steering Kugelumlauf-Servolenkung f
  ball-and-nut steering Kugelumlauflenkung f
  ball-and-socket gear change Kugelschaltung f *(beim Schaltgetriebe)*
  ball-and-socket joint Kugelgelenk n
  ball-and-socket shift Kugelschaltung f *(beim Schaltgetriebe)*
  ball-and-socket type gear shifting Kugelschaltung f
  ball bearing Kugellager n
  ball bearing bush Kugellagerbuchse f
  ball bearing cover Kugellagerdeckel m
  ball bearing protection Kugellagerschutz m

ball cup Kugelpfanne f, Kugelschale f
ball flange Kugelflansch m
ball freewheeling Kugelfreilauf m
ball gear shift Kugelschaltung f *(beim Schaltgetriebe)*
ball head Kugelkopf m
ball head screw Kugelkopfschraube f
ball housing Kugelgehäuse n
ball jet Kugeldüse f
ball joint Kugelgelenk n, Kugelzapfen m
ball joint socket Kugelgelenkgehäuse n
ball pin Kugelbolzen m
ball plate Kugelplatte f
ball race Laufring m, Kugellagerring m, Kugelring m
ball retainer Kugelkäfig m
ball seat nut Kugelbundmutter f
ball socket Kugelpfanne f, Kugelkopfpfanne f, Kugelbolzenpfanne f
ball thrust bearing Kugeldrucklager n
ball valve Kugelventil n
bearing ball Lagerkugel f
joint ball Gelenkkugel f
oblique ball bearing Schrägkugellager n
recirculating ball steering Kugelumlauflenkung f
row ball bearing Rillenkugellager n
self-aligning ball bearing Pendelkugellager n
shoulder ball bearing Schulterkugellager n
valve ball Ventilkugel f
balloon Ballon m
balloon spring Ballonfeder f
balloon tire Ballonreifen m
ballooning Bildung f ungewischter Flächen im Wischfeld *(beim Scheibenwischer)*
banana plug Bananenstecker m
band Band n, Bandage f
band brake Bandbremse f
band brake lining Bandbremsbelag m
band disk Bandscheibe f
band lock Bandageschloß n
clincher band Wulstband n
cover band Spannband n, Verschlußband n

cover band assembly   Spannband n
draw band   Zugband n
earth band   Masseband n
spring steel band   Federstahlband n
wide-band headlamp   Breitbandscheinwerfer m
bandwidth   Bandbreite f
banjo axle   Banjoachse f
bar   Stab m, Stange f
  bar filter   Stabfilter m,n
  bar spring   Stabfeder f
  antiroll bar   Querstabilisator m, Stabilisator m, Drehstab m
  compressed air bar   Druckluftleiste f
  connecting bar   Verbindungsstange f
  connector bar   Polbrücke f
  frame side bar   Längsträger m, Rahmenholm m
  gear-shift bar   Schaltstange f
  handle bar flash-light   Lenkerblinkleuchte f
  locking bar   Sicherungsriegel m
  motorcycle handle-bar end   Kraftradlenkerende n
  screen bar   Gitterstab m
  stop bar   Arretierstange f
  terminal bar   Polbrücke f
  torsion bar   Drehstabfeder f
  torsion bar spring   Drehstabfeder f
  tow-bar   Abschleppstange f
  turning bar   Durchdrehstab m
  wiper bar   Wischerstange f
barrel   Zylinder m; Lufttrichter m, Zerstäuber m *(beim Vergaser)*
  barrel bearing   Tonnenlager m
  barrel screw   Faßschraube f
  pinion barrel   Ritzelhülse f
  spark plug barrel   Kerzengehäuse n
barrier layer   Sperrschicht f
base   Sockel m
  base plate   Grundplatte f
  base rail   Unterlagschiene f
  base rim   Grundfelge f
  base spring   Unterlagfeder f
  axle base   Achsabstand m

coil base   Spulenschalter m
contact-breaker base   Unterbrecherhalteplatte f,
Unterbrechertragplatte f, Kontaktplatte f
crankshaft bearing base   Kurbelwellenlagerbock m
drop-base rim   Tiefbettfelge f
flat-base rim   Flachbettfelge f
lamp base   Glühlampensockel m
long wheel base   großer Radabstand m
rubber element base   Gummielementsockel m
switch base   Schalterkonsole f
well-base rim   Tiefbettfelge f
wheel base   Radstand m, Achsabstand m
wood base   Holzunterlage f
basic structure   Fahrwerk n *(ohne Motor)*, Unterbau m, Chassis n
bath   Bad n
  oil bath air cleaner   Ölbadluftfilter n,m
  oil bath air filter   Ölbadluftfilter n,m
battery   Batterie f, Akkumulator m
  battery-and-coil ignition   Sammlerzündung f
  battery body   Batteriegehäuse n
  battery box   Batteriegefäß n, Batteriekasten m
  battery box support   Batteriekastenträger m
  battery cable   Batteriekabel n
  battery casing   Batteriegehäuse n
  battery cell   Batteriezelle f
  battery cell plug   Zellenstopfen m
  battery cell tester   Batteriezellenprüfer m
  battery change-over switch   Batterieumschalter m
  battery charge   Batterieladung f
  battery charge failure indicator   Ladekontrolle f
  battery charger   Batterieladegerät n
  battery charging condition indicator lamp
  Ladekontrolleuchte f
  battery coil ignition   Batteriezündung f
  battery connection   Batterieanschluß m
  battery current   Batteriestrom m
  battery discharge   Batterieentladung f
  battery earthing lead   Batteriemassekabel n
  battery earthing strap   Batteriemasseband n

battery frame Batterierahmen m
battery ignition Batteriezündung f
battery jump cable Starthilfekabel n
battery lead Batterieleitung f, Batteriekabel n
battery main switch Batteriehauptschalter m
battery mud Batterieschlamm m
battery pole Batteriepol m
battery quick charger Batterieschnelladegerät n
battery switch Batterieumschalter m
battery terminal Batterieklemme f, Batterieanschlußklemme f, Batteriepol m
battery tester Batterieprüfer m, Batteriemeßinstrument n
battery throw-over switch Batterieumschalter m
battery voltage Batteriespannung f
starting battery Anlaßbatterie f
bayonet Bajonett-
  bayonet cap Bajonettsockel m, Bajonettverschluß m
  bayonet joint Bajonettanschluß m
  bayonet socket Bajonettfassung f
beacon Leuchtfeuer n, Verkehrsampel f
  rotating flashing beacon Rundumkennleuchte f
bead Reifenwulst m, Wulst m, Verstärkungsrippe f; Schweißraupe f
  bead base Wulstsohle f
  bead breaker Wulstheber m
  bead chafing Wulstscheuern n
  bead core Wulstkern m
  bead heel Wulstferse f
  bead ledge Wulstsitz m
  bead sealing ring Wustdichtring m
  bead seat Wulstsitz m, Felgenschulter f, Schulter f
  bead seating Reifensitz m
  bead seat radius Harnfußradius m, Schulterradius m
  bead seat width Schulterbreite f
  bead separation Wulstablösung f
  bead unseating Wulstabgleiten n, Abdrücken n des Wulstes vom Felgenhorn
  rim with taper bead seats Schrägschulterfelge f
  rubber bead Gummidichtung f
beam Balken m, Lichtstrahl m, Träger m

beam aim   Scheinwerfereinstellung f
beam axle   Starrachse f
beam deflector switch   Abblendschalter m
beam of light   Lichtstrahl m
cross beam   Querträger m, Riegel m
dipped beam   Abblendlicht n
distance beam headlight   Weitstrahler m
full (high) beam   Fernlicht n
long-distance beam headlight   Weitstrahler m
low beam   Abblendlicht n
lower passing beam   Abblendlicht n
main beam   Fernlicht n, aufgeblendetes Licht n
main beam indicator   Fernlichtkontrollampe f
main beam warning lamp   Fernlichtkontrollampe f
partial main beam   Teilfernlicht n
passing beam   Abblendlicht n
rigid beam axle   Starrachse f
sealed beam headlamp   Sealed-beam-Scheinwerfer m *(am. für ein
optisches System, bei dem Lichtaustrittscheibe, Reflektor und
Glühlampe eine Einheit sind)*
supporting beam   Tragbalken m
switch to the main beam/to   aufblenden
traffic beam   Abblendlicht n
bearing   Lager n, Lagerstelle f, Lagerung f
  bearing angle   Lagerwinkel m
  bearing arm   Lagerarm m
  bearing ball   Lagerkugel f
  bearing body   Lagerkörper m
  bearing bolt   Lagerbolzen m
  bearing bow   Lagerbügel m
  bearing bracket   Lagerstütze f, Tragbock m, Lagerbock m
  bearing brass   Lagerschale f
  bearing bush[ing]   Lagerschale f, Lagerbüchse f, Lagerbuchse f
  bearing cap   Lagerkappe f
  bearing casing   Lagergehäuse n
  bearing cover   Lagerdeckel m
  bearing disk   Lagerscheibe f
  bearing flange   Lagerflansch m
  bearing foot   Lagerfuß m

bearing liner   Lagermetall n
bearing load   Lagerbelastung f
bearing neck   Lagerzapfen m
bearing needle   Lagernadel f
bearing pin   Lagerzapfen m, Lagerstift m
bearing play   Lagerluft f, Spiel n
bearing rail   Tragschiene f
bearing ring   Lagerring m
bearing roller   Wälzzylinder m
bearing shell   Lagerschale f
bearing sleeve   Lagerhülse f
bearing tube   Lagerrohr n
axial bearing   Axiallager n
axle bearing   Achslager n
ball bearing   Kugellager n
ball bearing protection   Kugellagerschutz m
barrel bearing   Tonnenlager n
bevel roller bearing   Kegelrollenlager n
camshaft bearing   Steuerwellenlager n, Nockenwellenlager n
centering bearing   Zentrierlager n
clutch bearing   Ausrücklager n
connecting rod bearing   Pleuellager n
counter bearing   Gegenlager n
cover bearing   Deckellager n
crankshaft bearing   Kurbelwellenlager n
crankshaft bearing base   Kurbelwellen-Lagerbock m
disengaging bearing   Ausrücklagerung f
double bearing   Doppellager n
drive bearing   Antriebslager n
drive end bearing   Antriebslager n
driving bearing   Antriebslager n
end shield bearing   Schildlager n
flange bearing   Flanschlager n
footstep bearing   Fußlager n, Spurlager n
free-wheeling with ball bearings   Kugellagerfreilauf m
intermediate bearing   Zwischenlager n
main bearing   Hauptlager n
multi-point bearing   Mehrpunktlagerung f
neck of a bearing   Lagerhals m

    needle bearing   Nadellager n
    oblique ball bearing   Schrägkugellager n
    oiling bearing   Ringschmierlager n
    pivot bearing   Drehlager n
    plain bearing   Gleitlager n
    push bearing   Stößellager n
    radial clutch bearing   Kupplungsdrucklager n
    roller bearing   Wälzlager n, Rollenlager n
    round bearing   Rundlager n
    rubber bearing   Gummilager n
    self-aligning ball bearing   Pendelkugellager n
    self-aligning roller bearing   Pendelrollenlager n
    shaft bearing   Wellenlager n
    sliding bearing   Gleitlager n
    spherical cap bearing   Kalottenlager n
    spring bearing   Federlager n
    supplementary bearing   Zusatzlager n
    support bearing   Stützlager n
    throw-out bearing of the clutch   Kupplungsausrücklager n
    thrust bearing   Drucklager n
    wiper bearing   Scheibenwischerlager n
bellows   Balg m, Faltenbalg m
    air suction bellows   Luftansaugbalg m
    hand brake bellows   Handbremsbalg m
    leather bellows   Lederbalg m
    rubber bellows   Gummibalg m
    spring bellows   Federbalg m
belt/to   anschnallen, den Sicherheitsgurt anlegen
belt   Riemen m, Gurt m, Sitzgurt m
    belt drive   Riemenantrieb m
    belt fork   Riemengabel f
    belt guide   Riemenführung f
    belt latch mechanism   Gurtschloß n *(Sicherheitsgurt)*
    belt protection   Riemenschutz m
    belt pulley   Riemenscheibe f
    belt retractor   Aufrolleinrichtung f, Aufrollautomatik f
    belt slip   Keilriemenschlupf m
    belt strap   Gurtband n
    belt system   Gurtsystem n, Sitzgurtsystem n

belt tension   Riemenspannung f
belt tightener   Riemenspanner m
belt usage rate   Gurtbenutzungsrate f, Gurtanlegequote f
driving belt   Antriebsriemen m
fan belt   Ventilatorriemen m
flat belt   Flachriemen m
limit belt   Begrenzungsriemen m
narrow V-belt   Schmalkeilriemen m
reinforced belt structure   Gürtelkonstruktion f *(Reifen)*
safety belt   Sicherheitsgurt m
sliding belt   Gleitriemen m
toothed belt   Zahnriemen m
V-belt pulley   Keilriemenscheibe f, Keilriemenrolle f
belted tire   Gürtelreifen m
bench   Sitzbank f; Werkbank f
  test bench   Prüfstand m
bend/to   biegen
bend   Krümmung f, Biegung f, Kröpfung f, Kurve f
  exhaust bend   Austrittskrümmer m
  exhaust pipe bend   Auspuffrohrbogen m
bending   Biege-
  bending spring   Biegefeder f
  bending strength   Biegefestigkeit f
Bendix drive   Bendix-Getriebe n
Bendix-type starter   Schraubtriebanlasser m
bevel   Kegel m, Abschrägung f
  bevel drive pinion   Antriebskegelrad n
  bevel face ring   Fasenring m *(beim Kolben)*
  bevel gear   Kegelrad n
  bevel gear couple   Ritzel n mit Tellerrad
  bevel gear drive   Kegelradantrieb m
  bevel pinion   Kegelrad n, Kegelritzel n
  bevel roller bearing   Kegelrollenlager n
  axle bevel   Achskegelrad n
  differential axle-drive bevel gear   Differentialantriebskegelrad n
  differential bevel pinion   Ausgleichkegelrad n, kleines
  Differentialantriebskegelrad n
bias   Schräge f
  bias belted tire   Diagonalgürtelreifen m

bias ply tire Diagonalreifen m
bias spring Vorspannfeder f *(am Ventil)*
bimetal spring Bimetallfeder f
blade Flügel m, Schaufel f, Blatt n
　blade terminal Flachstecker m
　fan blade Ventilatorflügel m
　wiper blade Scheibenwischerblatt n
bleed/to ablassen, ausströmen lassen, entlüften, leck sein
bleeder Entlüftungs-
　bleeder hole Entlüftungsbohrung f
　bleeder valve Entlüftungsventil n
　air bleeder Bremsluftbohrung f
blend Gemisch n, Mischkraftstoff m
blind Jalousie f, Rollo n
block Block m, Klotz m, Blockierung f
　contact block Auflageklotz m
　cylinder block Zylinderblock m
　damping block Dämpfungsblock m
　dummy block Vorlegscheibe f
　radiator block Kühlerblock m
　rubber block Gummiblock m
　rubber brake block Gummibremsblock m
　sliding block Gleitstein m
　spacing block Distanzklotz m
　synchronizing block Synchronriegel m
　V-cylinder block Zylinderanordnung f in V-Form
board Brett n; Tafel f; gedruckte Schaltung f *(elektronisch)*
　foot board Trittbrett n, Trittplatte f
　fuse board Sicherungsbrett n
　instrument board Armaturenbrett n
　insulating board Isolierplatte f
　running board Trittbrett n
　splash board Auffangblech n
　terminal board Klemmleiste f, Klemmbrett n
　trailer board Anhängertafel f
　warning board Warntafel f
body Karosserie f, Aufbau m, Karkasse f *(Reifen)*
　body assembly line Karosseriemontageband n
　body chassis unit selbsttragende Karosserie f

body component Karosserieteil n
body deadener Antidröhnmaterial n
body front end Karosserievorbau m, Vorderbau m
body platform Bodenblech n, Bodengruppe f
body press Karosseriepresse f
body rear end Karosserieheck n, Hinterbau m
body shell Karosserie f, Rohbaukarosserie f
adjusting body Verstellkörper m
all-steel body Ganzstahlkarosserie f
automotive body Karosserie f
battery body Batteriegehäuse n
bearing body Lagerkörper m
car body Karosserie f, Aufbau m
carriage body screw Karrosserieschraube f
coil body Spulenkörper m
distribution body Verteilerkörper m
glow body Glühkörper m
insulating body Isolierkörper m
nozzle body Düsenkörper m
plug body Zündkerzenkörper m, Zündkerzenmantel m
pole body Polgehäuse n
starter body Anlassergehäuse n
unit-construction all-metal body selbsttragende
Ganzstahlkarosserie f
valve body Ventilkörper m
vehicle body Karosserie f
bolt Bolzen m, Schraube f *(mit Mutter)*
actuating bolt Schaltstift m
adjusting bolt Einstellbolzen m, Stellbolzen m
angle bolt Winkelbolzen m
ball head bolt Kugelbolzen m, Kugelkopfbolzen m
bearing bolt Lagerbolzen m
centering bolt Zentrierbolzen m
clamping bolt Klemmbolzen m
clinch bolt Nietbolzen m
collar bolt Bundbolzen m
conical bolt konischer Bolzen m
connecting bolt Anschlußbolzen m
contact bolt Kontaktbolzen m

control bolt   Schaltbolzen m
cylinder bolt   Zylinderbolzen m
eccentric bolt   Exzenterbolzen m
fastening bolt   Befestigungsbolzen m
fitted bolt   Paßbolzen m
fitting bolt   Anschlußbolzen m
fixing bolt   Befestigungsbolzen m
guide bolt   Führungsbolzen m, Lenkbolzen m
half bolt   Halbbolzen m
hexagon bolt   Sechskantbolzen m
hinge bolt   Scharnierbolzen m
holding bolt   Haltebolzen m
insert bolt   Steckbolzen m
joint bolt   Gelenkbolzen m
key bolt   Keilbolzen m
lock bolt   Sperriegel m
locking bolt   Steckriegel m, Verschlußbolzen m
pointed bolt   Spitzbolzen m
riveting bolt   Nietbolzen m
safety bolt   Sicherungsbolzen m
screw bolt   Schraubenbolzen m
sliding bolt   Schieberbolzen m
snug bolt   Nasenschraube f
spring bolt   Federbolzen m
spring center bolt   Federstift m
spring guide bolt   Federführungsbolzen m
stay bolt   Abstützbolzen m
steering stub bolt   Achsschenkelbolzen m
stop bolt   Abstellbolzen m, Rastbolzen m, Anschlagbolzen m, Arretierriegel m
supporting bolt   Tragbolzen m
taper bolt   Konusbolzen m
threaded bolt   Gewindebolzen m
threaded head bolt   Gewindekopfbolzen m
threaded hex-bolt   Sechskantgewindebolzen m
through bolt   Durchgangsbolzen m
welding bolt   Einschweißbolzen m
wheel bolt   Radbolzen m
bonnet   Motorhaube f

bonnet release Motorhaubenentriegelung f
engine bonnet Motorhaube f
book Buch n
  registration book Zulassungsschein m
boost Aufladung f
  boost pressure Ladedruck m
  boost pressure controller Ladedruckregler m
  boost pressure ga[u]ge Ladedrückanzeige f
  boost pressure limiter Ladedruckbegrenzer m
  boost pressure setting Ladedruckeinstellung f
boot Kofferraum m
  boot lid Kofferraumdeckel m
  boot lid support Kofferraumdeckelstütze f
  boot loading Kofferraumbelastung f
  luggage boot Kofferraum m
border Einfassung f, Umrandung f
  headlight border Scheinwerferzarge f
  rubber border Gummiabschluß m
  sheet metal border Blecheinfassung f
bore Bohrung f, Zylinderbohrung f
  bore-stroke ratio Hubverhältnis n, Hub-Bohrung-Verhältnis n
  connecting bore Anschlußbohrung f
bottom Boden m, unterster (tiefster) Teil m
  bottom dead center unterer Totpunkt m, UT
  bottom gear niedrigster Gang m, erster Gang m
  bottom level mark untere Markierung f *(beim Ölmeßstab)*
  bottom valve Bodenventil n
bow Bügel m, Dachbügel m, Dachspriegel m
  adjustment bow Einstellbügel m
  bearing bow Lagerbügel m
  clamping bow Klemmbügel m
  contact bow Kontaktbügel m
  driving bow Mitnehmerbügel m
  fastening bow Befestigungsbügel m
  folding bow Faltspriegel m
  roof bow Dachspriegel m
  safety bow Fangbügel m
  slide bow Gleitspriegel m
  snap bow Schnappbügel m

spring tension bow Federspannbügel m
stop bow Anschlagbügel m, Arretierbügel m
threaded bow Gewindebügel m
bowden cable (wire) Bowdenzug m, Bowdenkabel n, Seilzug m
bowl Schale f, Becher m, Schüssel f
 deep bowl reflector Tiefstrahler m
 filter bowl Filtertopf m, Filtergehäuse n
 funnel bowl Trichterschale f
 oil bowl Gehäuse n *(des Ölluftfilters)*
box Kiste f, Kasten m, Kabine f
 box section Kastenform f, Kastenprofil n
 box section member Kastenträger m
 battery box Batteriegefäß n, Batteriekasten m
 battery box support Batteriekastenträger m
 clutch coupling box Ausrückklaue f
 coil box Spulenkasten m
 connection box Anschlußkasten m
 control box Regler m *(für Lichtmaschinen)*
 distribution box Verteilerdose f
 flat plug box Flachsteckdose f
 fuse box Sicherungsdose f, Sicherungskasten m
 gear box Getriebe n, Wechselgetriebe n, Getriebegehäuse n
 gear box flange Getriebeflansch m
 gear-change box Wechselgetriebe n
 glove box lamp Handschuhkastenleuchte f
 grease box Schmierbüchse f
 junction box Anschlußdose f, Abzweigdose f
 link box Gelenkmuffe f
 low-pressure box Unterdruckdose f
 parallel box filter Parallelboxfilter m,n
 pole box Polgehäuse n
 protective box Schutzkasten m
 relay box Relaiskasten m
 steering box Lenkgetriebe n, Lenkgehäuse n
 steering box lever Lenkstockhebel m
 steering box support Lenkstockträger m
 steering box switch Lenkstockschalter m
 steering column box Lenkstockgehäuse n
 switch box Schaltkasten m, Schalterdose f

vacuum box Unterdruckdose f
brace Strebe f
bracing Verstrebung f, Versteifung f
  bracing ply Gürtellage f *(beim Reifen)*
  bracing tube Tragrohr n
bracket Träger m, Stütze f, Auflage f
  air cleaner bracket Luftansaugfilterstutzen m
  angle bracket Winkelhalter m
  angular bracket Winkelstütze f
  bearing bracket Lagerbock m, Lagerstütze f
  drive end bracket Antriebslagerschelle f
  fender bracket Kotflügelträger m
  guide bracket Führungssupport m
  headlight bracket Scheinwerferhalter m
  hinge bracket Scharnierbock m
  lamp bracket Lampenhalter m
  mirror bracket Rückspiegelhalter m
  mounting bracket Befestigungsbock m, Lagerbock m
  mudguard bracket Kotflügelhalter m
  pedal bracket Fußbetätigung f, Fußhebelsupport m,
Fußhebellagerbock m
  sliding bracket Gleitbock m
  spring slide bracket Federgleitbock m
  stop bracket Arretiersupport m
  welding bracket Anschweißbock m
brake Bremse f
  brake actuating lever Bremsbetätigungshebel m
  brake adjuster Bremsennachstellvorrichtung f,
Bremseneinstellvorrichtung f
  brake anchor plate Bremsankerplatte f, Bremsbackenhalter m,
Bremsträger m, Bremsbackenträger m
  brake application Bremsenbetätigung f
  brake balance Bremsenabstimmung f, Bremskraftverteilung f
  brake band Bremsband n
  brake bleeder valve Bremsentlüfterventil n
  brake booster Bremskraftverstärker m,
Bremskraftunterstützungsgerät n, Bremsgerät n
  brake bracket Bremsträger m
  brake buffer Bremsanschlag m

brake cable Bremsseil n, Bremszug m
brake cable assembly Bremsseilzug m *(als Baugruppe)*
brake caliper Bremssattel m, Bremszange f, Sattel m, Radsattel m
brake cam Bremsnocken m
brake cheek Bremsbacke f
brake circuit Bremskreis m
brake-circuit diagram Bremskreisschema n, Bremsschema n
brake collar Bremsband n
brake compensation Bremsausgleich m
brake compensator Bremsausgleich m, Bremslastausgleich m
brake conduit Bremsleitung f
brake control Betätigungseinrichtung f der Bremse
brake control rod Bremsstange f
brake coupling Bremskupplung f
brake cylinder Bremszylinder m
brake disk Bremsscheibe f
brake drum Bremstrommel f
brake equalizer Bremsausgleich m, Bremslastausgleich m
brake fluid Bremsflüssigkeit f
brake fluid level Bremsflüssigkeitsstand m
brake fluid reservoir Bremsflüssigkeitsbehälter m
brake force Bremskraft f
brake-force balance Bremskraftverteilung f
brake-force limiter Bremskraftbegrenzer m
brake-force proportioning device Bremskraftverteiler m
brake-force proportioning valve Bremskraftverteiler m
brake handle Bremshebelgriff m
brake horsepower Bremsleistung *(in PS)*
brake hose Bremsschlauch m
brake hub Bremsnabe f
brake imbalance fehlerhafte Bremsenabstimmung f
brake indicated power indizierte Leistung f, Innenleistung f
brake judder Bremsrubbeln n
brake lever Bremshebel m
brake lever connecting rod Bremszugstange f
brake line Bremsleitung f
brake lining Bremsbelag m, Bremsbackenbelag m
brake lining warning lamp Bremsbelagwarnleuchte f,
Bremsbelagkontrolleuchte f

brake lining wear Bremsbelagverschleiß m, Bremsbelagabnutzung f
brake lining wear indicator Bremsbelagverschleißanzeige f
brake linkage Bremsgestänge n
brake load Bremsenbelastung f
brake lockup Bremsenblockierung f
brake master cylinder Bremshauptzylinder m, Hauptbremszylinder m, Hauptzylinder m
brake operation Bremsenbetätigung f
brake operator lever Bremsbetätigungshebel m
brake outlet port Bremshauptzylinderausgang m, Hauptbremszyliderausgang m, Hauptzylinderausgang m
brake pad Bremsklotz m, Reibklotz m
brake pad wear indicator Bremsbelagverschleißanzeige f
brake pedal Bremspedal n
brake pedal clearance Bremspedalspiel n
brake pedal free travel Bremspedalfreigang m, Bremspedalspiel n
brake pedal linkage Bremspedalgestänge n
brake pedal play Bremspedalspiel n
brake pedal sealing cup Bremspedalmanschette f
brake pedal travel Bremspedalweg m
brake performance Bremsleistung f, Bremswirkung f
brake pipe Bremsrohrleitung f, Bremsleitung f
brake piston Bremskolben m
brake piston cup Bremsmanschette f
brake plate Bremsankerplatte f, Bremsträger m, Bremsbackenhalter m, Bremsbackenträger m
brake plate Bremsring m
brake plunger Bremskolben m
brake power Bremskraft f
brake power assist unit Bremskraftverstärker m, Bremskraftunterstützungsgerät n
brake power balance Bremskraftverteilung f
brake power compensation Bremslastausgleich m, Bremsausgleich m
brake power controller Bremskraftregler m
brake power distribution Bremskraftverteilung f
brake power equalizer Bremslastausgleich m, Bremsausgleich m
brake power limiter Bremskraftbegrenzer m

brake power limiting device Bremskraftbegrenzer m
brake power proportioning Bremskraftverteilung f
brake power proportioning device Bremskraftverteiler m
brake power proportioning valve Bremskraftverteilerventil n
brake power unit Bremskrafterzeuger m
brake pressure Vorratsbremsdruck m, Bremsdruck m, Bremskraft f
brake pressure control device Bremsdruckregler m, Bremskraftregler m
brake pressure control valve Bremsdruckregelventil n
brake pressure differential indication Bremsdruckdifferenzanzeige f
brake pressure reducer Bremskraftminderer m
brake pressure regulator Bremsdruckregler m, Bremskraftregler m
brake pressure tester Bremsdrucksprüfgerät n
brake pull Schiefziehen n der Bremsen, Bremsenschiefziehen n
brake pull rod Bremszugstange f
brake release Lösen n der Bremsen, Bremsauslösung f
brake release spring Bremsbackenrückholfeder f, Bremsbackenrückzugfeder f
brake resistance Bremswiderstand m
brake rod Bremsstange f
brake shoe Bremsklotz m, Bremsbacke f
brake shoe holder Bremsbackenhalter m
brake shoe lining Bremsbackenbelag m, Bremsbelag m
brake shoe pin bushing Bremsbackenlager n
brake shoe return spring Bremsbackenrückholfeder f, Bremsbackenrückzugfeder f
brake spring Bremsfeder f
brake squeal Bremsquietschen n, Quietschen n
brake stop Bremsanschlag m
brake strap Bremsband n
brake system Bremsanlage f
brake system indicator lamp Bremsanlagenkontrolleuchte f, Bremsenkontrolleuchte f
brake system integrity Unversehrtheit f der Bremsanlage f
brake system warning light Bremskreisausfallanzeige f, Bremanlagekontrolleuchte f
brake test Bremsenprüfung f, Bremsprobe f

brake torque Bremsmoment n
brake valve Bremsventil n
brake warning lamp Bremsenwarnleuchte f,
Bremsenkontrolleuchte f
brake wear Bremsbelagverschleiß m, Bremsklotzverschleiß m,
Bremsbelagabnutzung f
air-operated linkage brake Gestängedruckluftbremse f
armature brake Ankerbremse f
cable brake Seilzugbremse f
clutch brake Kupplungsbremse f
compressed-air brake Druckluftbremse f, pneumatische Bremse f
connecting rod of brake-lever Bremszugstange f
door brake Türbremse f
duplex brake Duplexbremse f
emergency brake Handbremse f, Feststellbremse f
exhaust engine brake Auspuffmotorbremse f
foot brake Fußbremse f, Betriebsbremse f
front-to-rear brake balance Bremskraftverteilung f *(Vorderachse
zu Hinterachse)*
front wheel brake Vorderradbremse f
gear brake Getriebebremse f
guide brake Führungsbremse f
hand brake Handbremse f
hand brake bellows Handbremsbalg m
hand brake cable Handbremsseil n
hand brake cylinder Handbremszylinder m
hand brake lever Handbremshebel m
hand brake servo Handbremsverstärker m
hand brake valve Handbremsventil n
hydraulic brake Öldruckbremse f
inside hand brake Innenhandbremse f
inside shoe brake Innenbackenbremse f
length of brake path Bremsweg m
linkage brake Gestängebremse f
loading brake Auflaufbremse f
magnetic brake Magnetbremse f
main brake cylinder Hauptbremszylinder m
master brake cylinder Hauptbremszylinder m
master cylinder of a brake Hauptbremszylinder m

motor brake   Motorbremse f
pedal brake   Fußbremse f, Fahrbremse f
power brake   Servobremse f
pull-on brake   Zughandbremse f
rubber brake block   Gummibremsblock m
servo-brake   Servobremse f
short-circuit brake   Kurzschlußbremse f
side-to-side brake balance   Bremsenabstimmung f an einer Achse,
Bremskraftverteilung f an einer Achse
single-chamber brake cylinder   Einkammerbremszylinder m
starter brake   Anlasserbremse f
trailer-brake valve   Anhängerbremsventil n
two-chamber brake cylinder   Zweikammerbremszylinder m
wheel brake cylinder   Radbremszylinder m
branch   Zweig m, Abzweigung f
connection branch   Anschlußstutzen m
break   Bruch m, Öffnung f, Kontaktabstand m
make-and-break contact   Umschaltkontakt m
make-before-break contact   unterbrechungsloser
Umschaltkontakt m
make-and-break ignition   Abreißzündung f
quick make-and-break switch   Momentschalter m
break-in/to   einfahren *(Motor)*
breaker   Unterbrecher m
breaker arm   Unterbrecherarm m
breaker cam   Unterbrechernocken m
breaker contact   Unterbrecherkontakt m
automatic circuit breaker   Selbstausschalter m
bead breaker   Wulstheber m *(für Reifen)*
circuit breaker   Schalter m, Ausschalter m
contact breaker   Unterbrecher m, Zündunterbrecher m
contact breaker adjustment   Unterbrechereinstellung f
contact breaker base   Kontaktplatte f, Unterbrecherhalteplatte f,
Unterbrechertragplatte f
contact breaker cable   Unterbrecherkabel n
contact breaker cam   Unterbrechernocken m, Kontaktnocken m
contact breaker gap   Unterbrecherkontaktabstand m,
Öffnungsspalt m *(Zündverteiler)*
contact breaker plate   Unterbrecherplatte f

contact breaker point Unterbrecherkontakt m
double-contact breaker Doppelunterbrecher m
ignition contact breaker Zündunterbrecher m
three-lever ignition contact breaker Dreihebelunterbrecher m
breather valve Entlüftungsventil n
breathing Entlüftung f
bridge Brücke f
  bearing bridge Lagerbrücke
  cable bridge Kabelbrücke
  connecting bridge Schaltbrücke f
  contact bridge Kontaktbrücke f
  measuring bridge Meßbrücke f
bridging Überbrückung f
  point bridging Überbrückung f des Elektrodenabstands
  (Zündkerze)
brush Kohlebürste f, Schleifkohle f, Kohle f
  carbon brush Kohlebürste f
  coal brush Kohlebürste f
  collector brush Stromabnehmer m
  graphite brush Graphitbürste f
  joint brush Abdichtbürste
  third-brush control Stromregelung f *(mittels Zusatzbürste an der
  Lichtmaschine)*
  third-brush dynamo stromregelnde Lichtmaschine f *(mit
  Zusatzbürste an der Lichtmaschine)*
buffer Dämpfer m, Puffer m
  damping buffer Dämpfungspuffer m
  stop buffer Anschlagpuffer m
  swinging-metal buffer Schwingmetallpuffer m
bulb Glühlampe f
  double-filament bulb Zweifadenglühlampe f, Biluxlampe f
  double-pole bulb Zweifadenglühlampe f
  flasher bulb Blinkerlampe f
  tubular bulb Röhrenglühlampe f
  two-filament bulb Zweifadenglühlampe f, Biluxlampe f
bumper Stoßfänger m, Stoßstange f, Federanschlag m,
  Anschlagpuffer m
  bumper bar Stoßfänger m, Stoßstange f
  bumper rubber strip Stoßstangengummileiste f

bumper strut Stoßfängerdämpfer m
corner bumper Eckenstoßfänger m
bush Büchse f, Lagerbüchse f *(s.a. bushing)*
  adjusting bush Einstellbüchse f
  ball bearing bush Kugellagerbuchse f
  bearing bush Lagerbuchse f, Lagerbüchse f, Lagerschale f
  centering bush Zentrierbüchse f
  collar bush Bundbüchse f
  control bush Steuerbüchse f
  coupling bush Kupplungsbüchse f
  driving bush Mitnehmerbüchse f
  free wheeling bush Freilaufbuchse f
  guide bush Führungsbüchse f
  insert bush Einsatzbüchse f
  oil lead-in bush Öleinführbüchse f
  press-in bush Eindrückbüchse f
  sinter bush Sinterbüchse f
  spacing bush Distanzbüchse f
  steering column bush Lenkstockbüchse f
  stop bush Anschlagbüchse f
  synchronous bush Synchronbüchse f
  tension bush Spannbüchse f
  threaded bush Gewindebüchse f
  toothed bush Zahnbüchse f
  welding bush Anschweißbüchse f *(s.a. bush)*
bushing Büchse f, Lagerbüchse f *(s.a. bush)*
  bearing bushing Lagerbuchse f
  cable bushing Kabeltülle f
  reverse idler gear bushing Rücklaufbuchse f
  rubber bushing Gummitülle f
butterfly Flügel-
  butterfly screw Flügelschraube f, Lappenschraube f
  butterfly valve Drosselklappe f
button Knopf m
  control button Schaltknopf m
  horn push button Horndruckknopf m
  notch button Rastknopf m
  push button Druckknopf m, Drucktaste, Knopf m
  push button switch Druckschalter m, Druckknopfschalter m

signal button Signalknopf m
signal push button Signaldruckknopf m
starter push button Anlaßschalter m, Anlasserdruckknopf m
switch button Schalterknopf m
bypass/to umgehen, umleiten, überbrücken
bypass Umleitung f, Bypass m, Nebenschluß m
  bypass bore Bypassbohrung f
  bypass governor Bypassregler m
  bypass thermostat Kurzschlußthermostat m *(zur Motorkühlung)*
  bypass valve Bypaßventil n, Umleitventil n, Sicherheitsventil n
  oil bypass valve Ölumleitventil n
cab Fahrerhaus n, Führerhaus n
  cab tilting system Fahrerhauskippvorrichtung f
  behind-engine cab Haubenfahrerhaus n, Haubenkabine f
cabinet Schrank m, Gehäuse n
  fuse cabinet Sicherungskasten m
cable Kabel n, Seil n
  cable bonder Leitungsverbinder m
  cable brake Seilzugbremse f
  cable bridge Kabelbrücke
  cable bushing Kabeltülle f
  cable clamp (clip) Drahtseilklemme f; Kabelklemme f, Kabelverbinder m
  cable conduit Seilhülle f; Kabelschutzrohr n
  cable connection Kabelanschluß m
  cable connector Leitungsverbinder m, Kabelverbinder m
  cable control Seilzugsteuerung f
  cable coupling Kabelkupplung f
  cable drum Kabeltrommel f
  cable duct Kabelkanal m
  cable eye Kabelöse f, Kabelschuh m
  cable grommet Kabeltülle f
  cable guide Kabelführung f
  cable head Kabelkopf m
  cable jacket Kabelhülle f, Kabelmantel m
  cable-operated seilzugbetätigt
  cable plug Kabelstecker m
  cable running Kabelverlegung f
  cable shoe Kabelschuh m

cable sleeve   Kabelverbinder m, Kabelhülse f
cable socket   Kabelschuh m
cable strand   Kabellitze f
cable strap   Kabelband n
cable terminal   Kabelklemme f
cable tree   Kabelbaum m
cable tube   Kabelrohr n
anti-interference cable   Entstörkabel n
battery cable   Batteriekabel n
connecting cable   Anschlußkabel n, Verbindungskabel n
contact breaker cable   Unterbrecherkabel n
hand brake cable   Handbremskabel n
ignition cable   Zündleitung f, Zündkabel n
ignition cable plug   Zündleitungsstecker m
radio-shielding cable   Entstörkabel n
starter cable   Anlasserkabel n
calibration   Eichung f, Einstellung f, Kalibrierung f
caliper   Bremssattel m, Sattel m
calorific value   Wärmewert m, Heizwert m
cam   Nocken m
cam carrier   Nockenwellenträger m
cam contour   Nockenprofil n
cam control   Nockensteuerung f
cam disk   Kurvenscheibe f, Nockenscheibe f
cam follower   Nockenstößel m
cam lever   Nockenhebel m
cam surface   Nockenflanke f, Nockenlaufbahn f
actuating cam   Betätigungsnocken m
clutch cam   Kupplungsnocken m, Kupplungsdaumen m
contact-breaker cam   Unterbrechernocken m, Kontaktnocken m
distributor cam   Verteilernocken m, Zündverteilernocken m
ignition cam   Unterbrechernocken m
stop cam   Abstellnocken m
camber   Sturz m, Radsturz m
camber angle   Radsturzwinkel m
camber rate   Sturzsteifigkeit f
camouflage headlamp   Tarnscheinwerfer m
camshaft   Nockenwelle f
camshaft arrangement   Nockenwellenanordnung f

camshaft bearing   Nockenwellenlager n, Steuerwellenlager n
camshaft belt   Steuerriemen m
camshaft casing   Nockenwellengehäuse n
camshaft chain   Steuerkette f
camshaft cover   Nockenwellendeckel m, Nockenwellenabdeckung f
camshaft drive   Nockenwellenantrieb m
camshaft drive chain   Nockenwellenantriebskette f, Steuerkette f
camshaft gear wheel   Nockenwellenrad n
camshaft housing   Nockenwellengehäuse n
camshaft sprocket   Nockenwellenkettenrad n
camshaft thrust bearing   Nockenwellenaxialsicherung f
camshaft timing   Nockenwellensteuerung f
camshaft timing gear   Nockenwellenantriebsrad n
overhead camshaft   obenliegende Nockenwelle f
cantilever   Konsole f, Ausleger m
cantilever spring   Auslegerfeder f
cantilever suspension   Zentralfederung f, Einzelfederaufhängung f
(beim Motorrad)
motor cantilever   Motorkonsole f
cap   Kappe f, Deckel m, Verschlußdeckel m
cap nut   Kappenmutter f, Hutmutter f
cap screw   Überwurfschraube f
bayonet cap   Bajonettsockel m
bearing cap   Lagerkappe f
closing cap   Abschlußkappe f
dimming cap   Abblendkappe f
distributor cap   Verteilerkappe f *(Zündverteiler)*
dust cap   Staubkappe f, Staubdeckel m
dust-proof cap   Staubdeckel m, Staubkappe f
end cap   Abdeckkappe f
fuel filler cap   Kraftstoffüllstutzen m
hub cap   Nabendeckel m, Radkappe m
joint covering cap   Abdeckkappe f
lamp cap   Glühlampensockel m
locking cap   Verschlußkappe f
petrol cap   Tankverschluß m
protective cap   Schutzkappe f
radiator filler cap   Kühlerdeckel m
screw cap   Gewindesockel m, Gewindestopfen m

spherical cap   Kalotte f, Kugelkalotte f
spherical cap bearing   Kalottenlager n
vent cap   Entlüftungskappe f
capacitor   Kondensator m
  capacitor case   Kondensatorgehäuse n
  capacitor flasher unit   Kondensatorblinkgeber m
  capacitor leads   Kondensatordurchführung f
  anti-interference capacitor   Entstörkondensator m
  ignition capacitor   Zündkondensator m
  radio-shielding capacitor   Entstörkondensator m
  tuned capacitor   Abstimmkondensator m
capacity   Kapazität f *(auch elektrische Kapazität)*,
  Fassungsvermögen n, Füllmenge f, Hubraum m
  capacity test   Kapazitätsprüfung f
  engine [volume] capacity   Hubraum m, Hubvolumen n
  load capacity   Belastungsfähigkeit f
  overload capacity   Überlastbarkeit f
capillary   Kapillar-
  capillary steel tube   Kapillarstahlrohr n
car   Wagen m, Auto n, Fahrzeug n
  car body   Karosserie f, Aufbau m
  car-door electric contact   Türkontakt m
  car range   Baureihe f *(Kraftfahrzeuge)*
  car type   Fahrzeugtyp m, Wagentyp m
  cross-country car   Geländewagen m
  motor car   Personenkraftwagen m, Pkw m
  under-car aerial   Chassisantenne f
carbon   Kohlenstoff m, Kohlestab m, Kohleelektrode f
  carbon brush   Kohlebürste f
  contact carbon   Kontaktkohle f
carburation   Gemischaufbereitung f, Vergasung f, Zerstäubung f
  carburation chamber   Mischkammer f
carburetor   Vergaser m
  carburetor adjustment   Vergasereinstellung f
  carburetor air heater   Ansaugluftvorwärmer m,
  Vergaserluftvorwärmer m, Vergaserluftheizung f
  carburetor assembly   Vergaseranlage f, "Vergaserbatterie" f
  carburetor body   Vergaserkörper m, Vergasergehäuse n
  carburetor bowl   Schwimmerkammer f

carburetor cable   Vergaserseilzug m, Vergaserzug m
carburetor control linkage   Vergasergestänge n
carburetor de-icing   Vergaserenteisung f
carburetor engine   Vergasermotor m
carburetor float   Vergaserschwimmer m, Schwimmer m
carburetor housing   Vergasergehäuse n, Vergaserkörper m
carburetor icing   Vergaservereisung f
carburetor jet   Vergaserdüse f
carburetor linkage   Vergasergestänge n
carburetor power jet   Anreicherungsdüse f
carburetor setting   Vergasereinstellung f
carburetor synchronization   Vergaserabstimmung f,
Vergasersynchronisierung f
**carburetor synchronization
device**   Vergaser-Synchrontestgerät n, Synchrontestgerät n
carburetor system   Vergaseranlage f, Vergasersystem n
carburetor throat   Vergaserlufttrichter m, Lufttrichter m
carburetor throttle   Vergaserdrossel f, Drosselklappe f
carburetor tuning   Vergasereinstellung f
carburetor venturi   Vergaserlufttrichter m, Lufttrichter m,
Zerstäuber m
cross-draught carburetor   Flachstromvergaser m
double jet carburetor   Doppeldüsenvergaser m
downdraught carburetor   Fallstromvergaser m
exhaust-jacket carburetor   Vergaser m mit Auspuffheizmantel m
heavy-oil carburetor   Schwerölvergaser m
horizontal draught carburetor   Flachstromvergaser m
jet carburetor   Düsenvergaser m, Zerstäubervergaser m
multijet carburetor   Mehrdüsenvergaser m
starting carburetor   Anlaßvergaser m, Startvergaser m
vacuum-controlled carburetor   Vergaser m mit
Unterdrucksteuerung f
carburetter, carburettor   s. carburetor
carcass   Karkasse f, Reifenkarkasse f
cardan   Kardan-
cardan axle   Kardanwelle f, Gelenkwelle f
cardan joint   Kardangelenk n
cardan shaft   Kardanwelle f
carriage   Fahrgestell n, Wagengestell n

carriage body   Karosserie f
carriage ga[u]ge   Wagenlehre f
carrier   Träger m, Gehäuse n
  carrier drum   Trägertrommel f
  carrier guide   Trägerführung f
  carrier plate   Mitnehmerplatte f
  carrier tube   Trägerrohr n
  differential carrier   Hinterachsbrücke f, Hinterachsträger m,
  Differentiallagerkasten m, Differentialgehäusevorderteil n,
  Ritzelträger m am Differential
  fuse carrier   Sicherungsträger m
  roof top carrier   Dachgepäckträger m
  sliding carrier   Schieberrahmen m
carrying   Trag-
  carrying axle   Tragachse f
  carrying capacity   Belastungsfähigkeit f, Tragfähigkeit f,
  Ladefähigkeit f
  carrying cup   Tragschale f
cartridge   Patrone f, Einsatz m
  filter cartridge   Filterpatrone f, Filtereinsatz m
  oil filter cartridge   Ölfilterpatrone f
  starting cartridge   Startpatrone f
case   Gehäuse n, Kasten m
  capacitor case   Kondensatorgehäuse n
  differential case (casing)   Ausgleichgehäuse n,
  Differentialgehäuse n
  distributor case   Verteilergehäuse n
  steering gear case   Lenkgehäuse n
  switch case   Schaltergehäuse n
  transfer case   Verteilergetriebe n
casing   Umhüllung f, Gehäuse n; Karkasse f, Reifenkarkasse f
  casing break up   Karkassenbruch m
  casing cover   Gehäusedeckel m
  casing ply   Karkassenlage f
  casing tube   Mantelrohr n *(an der Lenksäule)*
  battery casing   Batteriegehäuse n
  bearing casing   Lagergehäuse n
  headlamp casing   Scheinwerfergehäuse n
  pawl casing   Klinkengehäuse n

pole casing   Polgehäuse n
cast/to   gießen
  cast iron crankcase   Gußkurbelgehäuse n
  cast steel wheel   Stahlgußrad n
castle nut   Kronenmutter f
castor   Laufunterschied *(Vorlauf oder Nachlauf)*
  castor action   Vorlauf m; Nachlauf m
  castor angle   Vorlaufwinkel m, Nachlaufwinkel m
  castor offset   Vorlaufstrecke f, Nachlaufstrecke f
  castor trail   Vorlauf m, Nachlauf m
catalyst   Katalysator m
  catalyst charge   Katalysatorfüllung f
  catalyst degradation   Katalysatorschädigung f
  catalyst deterioration   Zerstörung f des Katalysators
catalytic converter   Katalysator m
catch   Eingriff m, Mitnehmer m, Arretierung f, Sperre f, Raste f, Klinke f
  catch lever   Klinkenhebel m, Sperrhebel m
  catch plate   Fangplatte f
  door catch   Türanschlag m, Türnase f
  hood catch   Haubenhalter m
  oil catch ring   Ölauffangring m
  safety catch lever   Sicherungshebel m
catcher   Schnapper n
cell   Zelle f, Element n, Batteriezelle f
  accumulator cell   Sammlerzelle f
  battery cell   Batteriezelle f
  battery cell plug   Zellenstopfen m
cellular radiator   Lamellenkühler m
center   Mittelpunkt m *(GB: centre)*
  center arm rest   Mittelarmstütze f, Mittelarmlehne f
  center bearing   Mittellager n
  center disk   Mittelscheibe f
  center electrode   Mittelelektrode f
  center member   Nabenteil n, Radnabenteil n
  center of attack   Seitenwindangriffspunkt m
  center of gravity   Schwerpunkt m
  center of gravity displacement   Schwerpunktverlagerung f, Schwerpunktverschiebung f

center pillar   Mittelwandsäule f, Mittelpfosten m, Mittelsäule f
center tube   Zentralrohr n
center-tubular chassis   Mittelrohrrahmen m,
Zentralrohrrahmen m
contact center   Auflagepunkt m *(Reifen)*
dead-center ignition   Totpunktzündung f
drop-center rim   Tiefbettfelge f
top dead center   oberer Totpunkt m
wheel center   Schüssel f *(Rad)*
centering   Zentrier-
  centering bearing   Zentrierlager n
  centering bolt   Zentrierbolzen m
  centering device   Zentriervorrichtung f
  centering disk   Zentrierscheibe f
  centering flange   Zentrierflansch m
central   Mittel-, Zentral-
  central chassis lubrication   Zentralschmierung f
  central electrode of spark plug   Kerzenstift m
  central greasing   Zentralschmierung f
  central locking   Zentralverschluß m, Zentralverriegelung f
  central tube frame   Mittelrohrrahmen m
centrifugal   Flieh[kraft]-, Zentrifugal-
  centrifugal advance   Fliehkraftverstellung f
  centrifugal clutch   Fliehkraftkupplung f
  centrifugal force   Fliehkraft f
  centrifugal governor   Fliehkraftregler m
  centrifugal mass   Schwungmasse f
  centrifugal spark advance   Fliehkraftzündversteller m,
Fliehkraftverstellung f, Fliehkraftversteller m
  centrifugal starter   Fliehkraftanlasser m
  centrifugal switch   Fliehkraftschalter m
  centrifugal timer   Fliehkraftversteller m
  centrifugal timing device   Fliehkraftversteller m
  centrifugal weight   Fliehgewicht n, Fliehkraftgewicht n
chafer   Wulstumlage f, Wulstband n *(Reifen)*
  chafer rubber   Wulstgummi n *(Reifen)*
chafing   Scheuern n *(Reifen)*
  chafing strip   Felgenband n
chain   Kette f

chain drive   Kettenantrieb m
chain guard   Kettenverdeck n, Kettenschutz m
chain link   Kettenglied n
chain lock   Kettenschloß n
chain pitch   Kettenteilung f
chain sprocket   Kettenrad n
chain tension roller   Kettenspannrolle f
chain tension wheel   Kettenspannrad n
chain tensioner (tightener)   Kettenspanner m
chain wheel   Kettenrad n
antiskid chain   Gleitschutzkette f
drive chain   Antriebskette f
driving chain wheel   Antriebskettenrad n
snow chain   Schneekette f
spur gear chain   Stirnradkette f
timing chain   Steuerkette f
transmission chain   Transmissionskette f
chamber   Kammer f
air chamber   Luftbehälter m, Luftkammer f
carburation chamber   Mischkammer f
float chamber   Schwimmergehäuse n *(Vergaser)*
pre-expansion chamber   Auspuffvorschalldämpfer m,
Vorschalldämpfer m
single-chamber brake cylinder   Einkammerbremszylinder m
two-chamber brake cylinder   Zweikammerbremszylinder m
change/to   wechseln, auswechseln, ändern
change down/to   herabschalten auf *(Gang)*
change gears/to   schalten *(Getriebe)*
change over/to   umschalten, schalten
change up/to   hochschalten *(Gang)*
change   Wechsel m
change filter   Wechselfilter m,n
change lever   Schalthebel m, Umschalthebel m
change of direction   Fahrtrichtungsänderung f
change-over jet   Übergangsdüse f
change-over shaft   Umschaltwelle f
change-over stop   Umschaltsperre f
change-over switch   Umschalter m, Drehrichtungsschalter m
change wheel   Wechselrad n

ball-and-socket gear change   Kugelschaltung f
battery change-over switch   Batterieumschalter m
dashboard gear change   Armaturenbrettschaltung f
gate change   Kulissenschaltung f
gear change   Gangschaltung f
gear-change box   Wechselgetriebe n
hydraulic gear change   Drucköl schaltung f
load change   Lastwechsel m
pneumatic change   Druckluftschaltung f
separate gear change   getrennte Schaltung f
speed-change gear   Gangschaltung f, Schaltgetriebe n
changing   Wechsel-
  gear changing   Gangwechsel m
channel   Fensterschacht m, Fensterscheibenführung f
characteristic   Kennzeichen n, Kennlinie f, Charakteristik f
  advance characteristic   Verstellinie f
  charging characteristic   Ladekennlinie f
  cornering characteristics   Kurvenlage f
  discharging characteristic   Entladekennlinie f
  dynamic characteristic   Arbeitskennlinie f
  idling characteristic   Leerlaufkennlinie f
  load characteristic   Belastungskennlinie f
  starter characteristic   Anlasserkennlinie f
charcoal   Aktivkohle f
charge/to   aufladen, laden; füllen
charge   Ladung f, Füllung f
  charge-air intercooling   Ladeluftkühlung f, Zwischenkühlung f
  charge air   Ladeluft f
  charge air cooler   Ladeluftkühler m
  charge air cooling   Ladeluftkühlung f
  charge air pressure   Ladeluftdruck m, Ladedruck m
  charge air tube   Ladeluftrohr n
  charge cycle   Ladungswechsel m, Gaswechsel m
  charge pressure   Ladedruck m, Ladungsdruck m
  charge stratifying   Ladungsschichtung f
  air charge tube   Luftladerohr n
  battery charge   Batterieaufladung f
  catalyst charge   Katalysatorfüllung f
  full charge   Volladung f

loss of charge   Ladeverlust m
charger   Ladegerät n
  battery charger   Batterieladegerät n
  turbo-charger   Turbolader m
charging   Aufladung f; Füllung f
  charging armature   Ladeanker m
  charging characteristic   Ladekennlinie f
  charging control   Ladekontrolle f
  charging current   Ladestrom m
  charging pressure adjusting device   Ladedruckeinsteller m
  charging pressure controller   Ladedruckregler m
  charging pressure ga[u]ge   Ladedruckmeßgerät n, Ladedruckkontrolle f
  charging pressure indicator   Ladedruckanzeigegerät n
  charging pressure limiting device   Ladedruckbegrenzer m
  charging rate   Ladegeschwindigkeit f
  charging voltage   Ladespannung f
  charging winding   Ladewicklung f
  battery charging   Laden n der Batterie
  dynamo charging light   Ladekontrollicht n
chassis   Chassis n, Fahrgestell n
  chassis and suspension tuning   Fahrwerkabstimmung f
  chassis beam   Längsträger m, Längsholm m
  chassis cross member   Rahmenquerträger m, Querträger m
  chassis dynamometer   Rollenprüfstand m
  chassis frame   Karosserierahmen m, Fahrgestellrahmen m, Chassisrahmen m
  chassis member   Längsträger m, Längsholm m
  chassis number   Fahrgestellnummer f
  chassis tuning   Fahrwerkabstimmung f
  central chassis lubrication   Zentralschmierung f
  forward control chassis   Frontlenker m
  low-built chassis   Niederrahmenfahrgestell n
  unitary construction of chassis and body   Aufbau m von Fahrgestell und Karosserie aus vereinheitlichten Teilen *(selbsttragende Karosserie)*
check   Prüfung f, Kontrolle f, Sperre f
  check disk   Anlaufscheibe f
  check strap   Türhalteriemen m

check valve Sperrventil n, Rückschlagventil n, Druckventil n
cheese head screw Zylinderschraube f
chock Unterlegkeil m, Bremsschutz m
  wheel chock Radkeil m
choke Drossel f, Luftklappe f, Starterklappe f, Choke m, Lufttrichter m
  choke blade Starterklappe f, Luftklappe f
  choke coil Drosselspule f
  choke control Starterzug m
  choke system Startvorrichtung f
  choke throttle lever Handgashebel m
  choke tube Drosselleitung f, Drosselkanal m, Lufttrichter m, Mischrohr n *(Vergaser)*, Venturieinsatz
  choke valve Starterklappe f
  choke warning light Luftklappenkontrollampe f
  screening choke Entstördrossel f
  suppressing choke Entstördrossel f
cigar lighter Zigarrenanzünder m
cigarette lighter Zigarettenanzünder m
circlip Nutring m, Sprengring m, Sicherungsring m, Seeger-Ring m
  circlip pliers Seeger-Zange f
  circlip sleeve Nutringmanschette f
circuit Kreislauf m, Stromkreis m, Kreis m, Schaltung f
  circuit breaker Unterbrecher m
  circuit diagram Schaltplan m
  automatic circuit-breaker Selbstausschalter m
  closed circuit Ruhestromkreis m, geschlossener Stromkreis m
  ignition circuit Zündschaltung f, Zündleitung f
  ignition circuit socket Zündleitungsstecker m
  low-potential circuit Niederspannungsstromkreis m
  low-voltage ignition circuit Niederspannungszündleitung f
  main circuit Hauptstromkreis m
  measuring circuit Meßkreis m
  multi-circuit switch Serienschalter m
  printed circuit gedruckte Schaltung f
  rotor circuit Läuferkreis m
  short circuit Kurzschluß m
  short-circuit armature Kurzschlußanker m
  short-circuit brake Kurzschlußbremse f

short-circuit rotor   Kurzschlußläufer m
short-circuit valve   Kurzschlußventil n
short-circuit voltage   Kurzschlußspannung f
shunt circuit   Nebenschlußstromkreis m
trigger circuit   Kippschaltung f
water return circuit   Wasserrückleitung f
circulation pump   Umwälzpumpe f
circumferential   umlaufend
circumferential groove   Längsrille f *(beim Reifen)*
circumferential joint   Ringdichtung f
clamp   Klammer f, Klemme f, Klemmstück n, Schelle f
clamp bearing   Klemmlager n
clamp bush   Klemmbuchse f
clamp collar   Klemmschelle f, Spannschelle f
clamp coupling   Klemmkupplung f
clamp disk   Klemmscheibe f
cable clamp   Kabelklemme f
final clamp   Abschlußklemme f
hose clamp   Schlauchbinder m
pipe clamp   Befestigungsschelle f
rim clamp   Felgenklammer f, Felgenschuh m
screw clamp   Schraubzwinge f
clamping   Klemm-
clamping bolt   Klemmbolzen m
clamping device   Klemmvorrichtung f, Aufspannvorrichtung f
clamping jaw   Klemmbacke f, Spannbacke f
clamping lever   Klemmhebel m
clamping screw   Klemmschraube f
clamping sleeve   Klemmuffe f
clamping strap   Spannband n
claw   Klaue f
claw clutch   Klauenkupplung f
claw coupling   Klauenkupplung f
claw joint   Klauengelenk n
claw ring   Klauenkranz n
claw wheel   Klauenrad n
coupling claw   Kupplungsklaue f
clean/to   reinigen
cleaner   Filter m,n

air cleaner bracket   Luftansaugfilterstutzen m,n
compressed-air cleaner   Druckluftfilter m,n
oil-wetted air cleaner   Naßluftfilter m,n
cleaning   Reinigung f
  cleaning agent   Reinigungsmittel n
clear vision screen   Klarsichtscheibe f
clearance   Spiel n, Luft f
  clearance circle   Wendekreis m
  clearance lamp   Begrenzungsleuchte f
  ground clearance   Bodenfreiheit f, Rahmenhöhe f
  road clearance   Bodenfreiheit f, Bodenabstand m
  valve clearance ga[u]ge   Ventilspiellehre f
clinch   Felgenhorn n
  clinch bolt   Nietbolzen m
clincher   Wulst m
  clincher band   Wulstband n
  clincher rim   Wulstfelge f
  clincher tire   Wulstreifen m
clip   Bügel m, Schelle f, Haltfeder f
  cable clip   Kabelverbinder m
  hose clip   Schlauchklemme f
  mounting clip   Befestigungsschelle f
  spring clip   Federbügel m, Federklemme f
clock   Uhr f
  motor car clock   Autouhr f
clockwise   im Uhrzeigersinn, rechtsdrehend
  anti-clockwise rotation   Linksdrehung f, Linkslauf m
  clockwise rotation   Rechtsdrehung f
clog/to   verstopfen
close/to   schließen, sperren, abdichten
closed   geschlossen
  closed circuit   Ruhestromkreis m, geschlossener Stromkreis m
  closed-circuit system   Ruhestromschaltung f
closing   Schließ-, Abschluß -
  closing angle   Abschlußwinkel m, Schließwinkel m
  closing cap   Abschlußkappe f
  closing element   Schließelement n
  closing screw   Abschlußschraube f
  contact closing   Kontaktschluß m

clothing hook   Kleiderhaken m
cloth lining   Stoffbelag m
clutch   Kupplung f
  clutch bearing   Ausrücklager n
  clutch brake   Kupplungsbremse f
  clutch cam   Kupplungsnocken m
  clutch case   Kupplungsgehäuse n
  clutch collar   Kupplungsmuffe f
  clutch cone   Kupplungskonus m, Kupplungskegel m
  clutch control valve   Kupplungsteuerventil n
  clutch coupling   Kupplungsgelenk n
  clutch coupling box   Ausrückklaue f
  clutch cover   Kupplungsdeckel m
  clutch disk   Kupplungsscheibe f
  clutch disk lining   Kupplungsbelag m
  clutch drive plate   Kupplungsmitnehmerscheibe f
  clutch drum   Kupplungstrommel f
  clutch housing   Kupplungsgehäuse n
  clutch hub   Kupplungsnabe f
  clutch lamella   Kupplungslamelle f
  clutch lever   Kupplungshebel m
  clutch lining   Kupplungsbelag m
  clutch of the starter   Freilaufgetriebe n des Anlassers
  clutch pedal   Kupplungspedal n
  clutch plate   Kupplungsscheibe f
  clutch presser plate   Kupplungsdruckplatte f
  clutch release lever   Ausrückhebel m
  clutch release sleeve   Ausrückmuffe f
  clutch release yoke   Ausrückgabel f
  clutch shaft   Kupplungsswelle f
  clutch spring   Kupplungsfeder f
  clutch weight   Kupplungsfliehgewicht n
  centrifugal clutch   Fliehkraftkupplung f
  disk clutch   Scheibenkupplung f, Reibscheibenkupplung f
  dog clutch   Klauenkupplung f
  dry multi-plate clutch   Mehrscheibentrockenkupplung f
  dry-plate clutch   Trocken[scheiben]kupplung f
  jaw-clutch   Klauenkupplung f
  lamella clutch   Lamellenkupplung f

multi-plate clutch   Lamellenkupplung f
multi-plate friction clutch   Lamellenrutschkupplung f
overload clutch   Überlastkupplung f
overrunning clutch   Freilaufkupplung f, Überholkupplung f
radial clutch bearing   Kupplungsdrucklager n
release the clutch pedal/to   einkuppeln
shoe clutch   Backenkupplung f
slipping clutch   Rutschkupplung f
step on the clutch pedal/to   auskuppeln, auf die Kupplung treten
throw-out bearing of the clutch   Kupplungsausrücklager n
coach   Omnibus m, Bus m
  coachbuilder   Karosseriehersteller m
  coachwork   Karosserie f, Wagenaufbau m
coal   Kohle f
  coal brush   Kohlenbürste f
coast/to   schieben, im Leerlauf m fahren
coasting   Schubbetrieb m
coat of paint   Lackschutz m
coating   Überzug m
code   Code m
  code insert   Codiereinsatz m
  code pin   Codierstift m
coefficient   Beiwert m, Koeffizient m
  coefficient of aerodynamic drag   Luftwiderstandsbeiwert m
  temperature coefficient   Temperaturkoeffizient m
coil   Spule f, Zündspule f
  coil base   Spulenschalter m
  coil body   Spulenkörper m
  coil ignition   Spulenzündung f, Batteriezündung f
  coil shell   Spulenkörper m
  coil spring   Schraubenfeder f
  battery coil ignition   Batteriezündung f
  battery-and-coil ignition   Sammlerzündung f
  distributor coil   Zündspule f
  exciting coil   Erregerspule f
  ignition coil   Zündspule f
  induction coil   Induktionsspule f
  starting coil   Anlaßspule f
cold   kalt

cold plug Zündkerze f mit hohem Wärmewert
cold start device Kaltstartvorrichtung f, Kaltstarteinrichtung f
cold-starting aid Kaltstarthilfe f
cold start valve Kaltstartventil n
collapsible faltbar
  collapsible top Faltdach n, Klappverdeck n
collar Manschette f, Muffe f, Band n, Büchse f
  collar bush Bundbüchse f
  collar screw Bundschraube f
  collar unit Bundmutter f
  clutch collar Kupplungsmuffe f
  thrust collar Druckring m
collect/to sammeln, zusammenfassen
collecting pipe Sammelrohr n
collector Kollektor m, Sammelleitung f, Krümmer m
  collector brush Stromabnehmer m
  collector end shield bearing Kollektorlagerschild m
  ignition impulse collector Zündimpulsabnehmer m
color Farbe f *(GB: colour)*
  colo[u]r code Kennfarbe f
  dipping colo[u]r Tauchfarbe f
column Säule f
  column gear shift Lenkradschaltung f
  column shift Lenkradschaltung f
  steering column Lenksäule f
  steering column box Lenkstockgehäuse n
combination Kombination f, Verbindung f; Gespann n,
  Seitenwagenmaschine f *(Motorrad)*
  combination pliers Kombizange f
combined kombiniert
  combined flash-tail lamp Kombinationsleuchte f
  (Blink-Schluß-Leuchte)
  combined mat Kombimatte f
combustible brennbar, zündfähig
  combustible mixture brennbares Gemisch n
combustion Verbrennung f
  combustion chamber Brennraum m, Verbrennungsraum m,
  Verbrennungskammer f
  combustion chamber deposit Brennraumablagerung f,

Ablagerung f im Verbrennungsraum
combustion chamber shape   Brennraumform f,
Verbrennungsraumform f
combustion deposits   Verbrennungsrückstände f
combustion energy   Verbrennungsenergie f
combustion engine   Wärmekraftmaschine f,
Verbrennungskraftmaschine f, Verbrennungsmotor m
combustion knock   Klopfen n
combustion peak pressure   maximaler Verbrennungsdruck m
combustion pressure   Verbrennungsdruck m
combustion principle   Verbrennungsverfahren n
combustion process   Verbrennungsablauf m,
Verbrennungsvorgang m
combustion residues   Verbrennungsrückstände f
combustion space   Verbrennungsraum m
combustion space deposit   Brennraumablagerung f, Ablagerung f
im Verbrennungsraum
combustion turbulence   Verbrennungsdrall m, Sekundärdrall m
internal combustion engine   Verbrennungsmotor m
comfort   Fahrtkomfort m
commutator   Kollektor m *(Elektromotor)*
commutator end bearing   Kollektorlager n
compact car   Kompaktwagen m, PKW m der US-Kompaktklasse
compartment   Kabine f, Fach n
glove compartment   Handschuhfach n
passenger compartment   Fahrgastraum m
compensating   Ausgleich-
compensating disk   Ausgleichscheibe f
compensating jet   Ausgleichsdüse f
compensating jet carburetor   Vergaser m mit Luftkorrekturdüse f
compensating tank   Ausgleichbehälter m
compensating tappet   Ausgleichstößel m
compensation   Kompensation f, Ausgleich m
load compensation   Belastungsausgleich m, Lastausgleich m
compound   Gemisch n, Masse f; zusammengesetzt, Verbund-
compound carburetor   Registervergaser m
compound engine   Verbundmotor m
compound suspension   Verbundfederung f
insulating compound   Isoliermasse f

compressed komprimiert
  compressed-air bar Druckluftleiste f
  compressed-air brake Druckluftbremse f, pneumatische Bremse f
  compressed-air cleaner Druckluftfilter n,m
  compressed air tube Druckluftschlauch m
compression Verdichtung f, Kompression f *(Motor)*; Einfederung f, Radeinfederung f *(Radaufhängung)*
  compression cycle Verdichtungstakt m
  compression ga[u]ge Kompressionsdruckmesser m, Kompressionsmesser m
  compression ignition Selbstzündung f, Eigenzündung f, Kompressionszündung f, Verdichtungszündung f
  compression ignition engine Dieselmotor m
  compression pressure Kompressionsdruck m, Verdichtungsdruck m
  compression ratio Verdichtungsverhältnis n, Kompressionsverhältnis n
  compression ring Verdichtungsring m, Kompressionsring m
  compression spring Druckfeder f
  compression stroke Kompressionshub m, Verdichtungshub m, Verdichtungstakt m
  compression strut Schubstange f, Schubstrebe f
  compression tester Kompressionsprüfer m, Kompressionsdruckmesser m, Kompressionsprüfgerät n
compressor Verdichter m, Kompressor m; Aufladegebläse n
  compressor housing Verdichtergehäuse n, Gebläsegehäuse n
  compressor inlet Verdichtereinlauf m, Gebläseeinlauf m, Lufteintritt m
  compressor outlet Verdichterauslauf m, Gebläseauslauf m, Luftaustritt m
  compressor wheel Verdichterrad n, Gebläserad n
concealed headlamp versenkter Scheinwerfer m
condenser Kondensator m
condition Zustand m, Beschaffenheit f
  liner condition Zylinderlaufbild n
  wear condition of liner Zylinderlaufbild n, Zylindertragbild n
conditioner Klimatisierer m
  air conditioner Klimaanlage f
conductor Leiter m

conductor rail Stromschiene f
conduit Isolierrohr n
  cable conduit Kabelschutzrohr n
cone Kegel m
  clutch cone Kupplungskonus m
  sliding cone Schiebekonus m
  synchronizing cone Synchronkegel m
  valve cone Ventilkegel m
connect/to anschließen; schalten; verbinden
  connect in parallel/to parallelschalten
  connect in series/to hintereinanderschalten
connecting Verbindungs-, Anschluß -
  connecting bar Verbindungsstange f
  connecting bore Anschlußbohrung f
  connecting bridge Schaltbrücke f
  connecting cable Anschlußkabel n, Verbindungskabel n
  connecting cover Anschlußdeckel m
  connecting elbow Anschlußkrümmer m
  connecting line Verbindungsleitung f
  connecting nut Anschlußmutter f
  connecting piece Verbindungsstück n, Anschlußteil n, Anschlußstück n
  connecting pipe Anschlußrohr n, Verbindungsstutzen m, Verschraubstutzen m
  connecting pole Anschlußpol m
  connecting rail Verbindungsschiene f
  connecting rod Pleuel n, Schubstange f, Pleuelstange f, Verbindungsstange f
  connecting rod bush Pleuelbüchse f
  connecting rod eye Pleuelauge n
  connecting rod shaft Pleuelschaft m
  connecting screw Verbindungsschraube f, Anschlußschraube f
  connecting strut Verbindungsstrebe f
  connecting terminal Anschlußklemme f
  connecting tube Verbindungsrohr n
  connecting wire Verbindungsdraht m
  angle of connecting rod Pleuelstangenausschlag m
  flat connecting bush Flachsteckhülse f
connection Anschluß m, Verbindung f, Schaltung f, Stutzen m

connection branch Anschlußstutzen m
connection fitting Anschlußnippel m
connection line Kupplungsleitung f
air vent connection Entlüfterstutzen m
angular connection Winkelanschluß m
armature end-connection Spulenkopf m
battery connection Batterieanschluß m
diagram of connections Schaltbild n
double connection Doppelstutzen m
earth connection Masseanschluß m
line connection Leitungsanschluß m
pipe connection Rohranschluß m
plug-and-socket connection Steckverbindung f
primary connection Primäranschluß m
reduction connection Reduzierstutzen m
ring connection Ringanschluß m
screw connection Schraubanschluß m, Schraubverbindung f
shunt connection Nebenschlußschaltung f
connector Anschlußteil n, Steckverbindung f
connector lug Kabelschuh m
connector socket Anhängersteckdose f, Kupplungssteckdose f
cable connector Leitungsverbinder m, Kabelverbinder m
snap connector Leitungsschnappverbindung f
console Konsole f
constant konstant, gleichbleibend
constant mesh gearbox Schubklauengetriebe n
constant vacuum carburetor Gleichdruckvergaser m,
Schiebervergaser m
constant velocity joint Gleichlaufgelenk n
constant velocity universal joint Gleichlaufgelenk n
construction Konstruktion f, Ausführung f, Bauart f
frameless construction selbsttragende Konstruktion f, rahmenlose
Bauart f
integral frame construction rahmenlose Bauweise f, voll
selbsttragender Aufbau m
tubular construction Rohrrahmenkonstruktion f
unit construction all-metal body selbsttragende
Ganzstahlkarosserie f
consumption Verbrauch m

fuel-consumption ga[u]ge   Kraftstoffverbrauchsmesser m
level road fuel consumption   Kraftstoffnormverbrauch m
power consumption   Leistungsaufnahme f, Leistungsbedarf m,
Leistungsverbrauch m
contact   Kontakt m; Kontaktstück n
    contact angle   Kontaktwinkel m
    contact bracket   Kontaktträger m
    contact breaker   Unterbrecher m, Zündunterbrecher m
    contact breaker adjustment   Unterbrechereinstellung f
    contact breaker arm   Unterbrecherhebel m
    contact breaker base   Kontaktplatte f
    contact breaker cable   Unterbrecherkabel n
    contact breaker cam   Unterbrechernocken m
    contact breaker gap   Unterbrecherkontaktabstand m,
Öffnungsspalt m *(Zündverteiler)*
    contact breaker plate   Unterbrecherplatte f
    contact breaker point   Unterbrecherkontakt m
    contact bridge   Kontaktbrücke f
    contact carbon   Kontaktkohle f
    contact center   Auflagepunkt m *(Reifen)*
    contact finger   Kontaktfinger m
    contact gap   Kontaktabstand m
    contact lever   Unterbrecherhebel m, Kontakthebel m
    contact piece   Kontaktstück n
    contact pin   Kontaktstift m
    contact plate   Kontaktplatte f, Aufschlagplatte f, Auflageplatte f,
Kontaktteller m
    contact plug   Steckereinsatz m
    contact rail   Stromschiene f, Kontaktschiene f
    contact resistance   Übergangswiderstand m
    contact ring   Auflagering m, Kontaktring m
    contact sleeve   Kontakthülse f
    contact spring   Kontaktfeder f
    contact support   Kontaktträger m
    distributor contact   Unterbrecherkontakt m
    door contact   Türkontakt m
    door contact interrupter   Türkontaktschalter m
    double-contact breaker   Doppelunterbrecher m
    ignition contact breaker   Zündunterbrecher m

ignition contact lever Zündkontakthebel m
plug contact Steckkontakt m, Stecker m
short-circuit contact Kurzschlußkontakt m
sliding contact Schleifkontakt m, Gleitkontakt m
contactor Impulsgeber m, Kontaktgeber m, Schaltschütz m
  contactor timer Schaltzeitregler m
  main contactor Hauptschütz m
container Behälter m
  anti-frost container Frostschutzbehälter m
  battery container Batteriebehälter m
  compressed-air container Druckluftbehälter m
  oil container Ölbehälter m, Ölfangbecken n
continuous kontinuierlich, Dauer-
  continuous-flow fuel injection Konstanteinspritzung f
  continuous fuel injection Konstanteinspritzung f
  continuous lubrication Dauerschmierung f
  continuous operation Dauerbetrieb m
  continuous rating Dauerleistung f
control Betätigung f; Betätigungseinrichtung f, Bedienteil n,
  Bedienungselement; Steuerung f; Regelung f
  control arm Querlenker m, Lenker m
  control box Regler m
  control bush Steuerbüchse f
  control button Schaltknopf m
  control cylinder Steuerzylinder m, Schaltzylinder m
  control element Steuerelement n, Schaltelement n
  control for headlamp adjustment Leuchtweitensteller m
  control ga[u]ge Kontrollehre f
  control housing Schaltgehäuse n
  control knob Betätigungsknopf m
  control lamp Kontrollglühbirne f, Kontrollampe f
  control lever Schalthebel m, Betätigungshebel m, Steuerhebel m
  control piston Steuerkolben m
  control plate Schalterplatte f, Steuerplatte f
  control rod Betätigungsstange f
  control slide Betätigungsschieber m
  control solenoid switch Steuermagnetschalter m
  control switch Steuerschalter m
  alternator control Wechselstromregelung f

cable control Seilzugsteuerung f
charging control Ladekontrolle f
directional control Spurführung f
direction-indicator control lamp Blinkerkontrolleuchte f
distant control Fernsteuerung f
drive-slip control Antrieb-Schlupf-Regelung f
field control Feldregelung f
flash control Blinkkontrolle f
forward control chassis Frontlenker m
fuel-air ratio control unit Gemischregler m
hand control Handschaltung f
hand control valve Handsteuerventil n
ignition control Zündversteller m, Zündzeitverstellung f
indicator control switch Winkerschalter m
knock control Klopfregelung f
longitudinal control arm Längslenker m
mixture control Gemischregelung f
oil pressure control switch Öldruckkontrollschalter m
remote control Fernsteuerung f
track control arm Spurstangenhebel m
vacuum timing control Unterdruckzünd[zeitpunkt]verstellung f
voltage control Spannungsregelung f
controlled gesteuert
  cable-controlled mit Seil[zug]steuerung ausgerüstet
  thermically controlled thermisch gesteuert
  vacuum-controlled carburetor Vergaser m mit
  Unterdrucksteuerung
control[l]ing armature Steueranker m
converter Umformer m, Wandler m *(z.B. beim Getriebe)*
  converter slip Wandlerschlupf m
  converter transmission Wandlerautomatik f, Wandlergetriebe n
convertible car Kabriolet n
coolant Kühlmittel n, Kühlflüssigkeit f, Kühlwasser n
  coolant drain plug Kühlmittelablaßschraube f,
  Wasserablaßschraube f
  coolant hose Kühlmittelschlauch m, Kühlwasserschlauch m,
  Wasserschlauch m
  coolant inlet connection Kühlwassereinlaufstutzen m
  coolant level Kühlmittelstand m, Kühlflüssigkeitsstand m,

Kühlwasserstand m, Wasserstand m
coolant level indicator Kühlmittelstandanzeige f,
Kühlflüssigkeitsstandanzeige f, Kühlwasserstandanzeige f,
Wasserstandanzeige f
coolant outlet connection Kühlwasserauslaufstutzen m
cooler Kühler m
 oil cooler Schmierölkühler, Ölkühler m
cooling Kühlung f
 cooling air Kühlluft f
 cooling-air stream Kühlluftstrom m
 cooling blower Kühlluftgebläse n
 cooling coil Kühlschlange f
 cooling element Kühlelement n
 cooling fin Kühlrippe f
 cooling liquid Kühlflüssigkeit f, Kühlwasser n, Kühlmittel n
 cooling liquid circulation Kühlflüssigkeitskreislauf m
 cooling pipe Kühlleitung f
 cooling rib Kühlrippe f
 cooling system Kühlsystem n, Kühlanlage f
 cooling unit Kühlaggregat n
 cooling water Kühlwasser n, Kühlflüssigkeit f, Kühlmittel n
 cooling water circulation Kühlwasserkreislauf m
 cooling water cycle Kühlwasserkreislauf m
copper Kupfer n
 copper gasket Kupferdichtung f
cord Seil n; Gewebe n *(beim Reifen)*
 cable cord Kabellitze f
core Kern m, Drahtader f
 core of cable Kabelader f, Ader f
 core plug Kernstopfen m
corner Ecke f, Eckblech n, Winkel m
 corner bumper Eckenstoßfänger m
 connecting corner Anschlußwinkel m
 metal corner Blechwinkel m
corner/to kurvenfahren
cornering Kurvenfahrt f
 cornering ability Kurvenfahrverhalten n
 cornering characteristics Kurvenlage f
 cornering force Seitenkraft f

cornering rate Kurvensteifigkeit f
cornering stability Kurvenstabilität f
correction Korrektur f, Verbesserung f
  correction air Ausgleichsluft f, Korrekturluft f
  air correction jet Ausgleichluftdüse f, Luftkorrekturdüse f
corrosion Korrosion f
  corrosion inhibitor Korrosionsschutzmittel n
  corrosion-proof korrosionsbeständig
  corrosion-protected korrosionsgeschützt
  corrosion resistance Korrosionsbeständigkeit f
  corrosion test Korrosionsprüfung f
  protection against corrosion Korrosionsschutz m
counter Zähler m, Zählwerk n; Gegen-
  counter bearing Gegenlager n
  counter gear Vorgelege n
  counter nut Kontermutter f
  revolution counter Drehzahlmesser m, Tourenzähler m
counterbalance/to auswuchten
countershaft Vorgelegewelle f, Vorgelege n
  countershaft gear Vorgelege n
  countershaft wheel Vorgelegerad n
countersunk screw Senkschraube f
counterweight Gegengewicht n
country Feld n, Land n
  cross-country car Geländewagen m
  cross-country properties Geländegängigkeit f
couple/to kuppeln
couple Paar n
  bevel gear couple Ritzel n mit Tellerrad
  gear couple Zahnradpaar n
coupling Kupplung f
  coupling bush Kupplungsbüchse f
  coupling claw Kupplungsklaue f
  coupling disk Kupplungsscheibe f
  coupling flange Kupplungsflansch m
  coupling lever Kupplungshebel m
  coupling plug Kupplungsstecker m
  coupling rod Kupplungsstange f, Kuppelstange f
  coupling segment Kupplungslamelle f

coupling shaft   Kupplungswelle f
coupling sleeve   Kupplungsmuffe f, Kupplungshülse f
coupling spur wheel   Kupplungsstirnrad n
coupling wheel   Kupplungsrad n
back-coupling   Rückkopplung f
claw coupling   Klauenkupplung f
clutch coupling box   Ausrückklaue f
hydraulic coupling   Flüssigkeitskupplung f
joint coupling   Gelenkkupplung f
tee coupling   T-Stutzen m
tow-bar coupling   Abschleppkupplung f
trailer coupling   Anhängerkupplung f
cover   Abdeckung f, Deckel m, Blende f, Haube f, Klappe f,
Abdeckklappe f, Hülle f
cover band   Spannband n, Verschlußband n
cover bearing   Deckellager n
cover plate   Deckplatte f, Deckblech n
cover strap   Abdeckband n
cover strip   Deckleiste f
air relief cover   Entlüfterdeckel m
bearing cover   Lagerdeckel m
casing cover   Gehäusedeckel m
cylinder cover   Zylinderdeckel m
drive cover   Antriebsdeckel m
filter cover   Filterdeckel m
gear-shift cover   Schaltdeckel m
governor cover   Reglerhaube f
housing cover   Gehäusedeckel m
lock cover   Verschlußdeckel m
protective cover   Schutzdeckel m
regulator cover   Reglerkappe f
seal cover   Abdichtdeckel m
covering   Abdeckung f, Verkleidung f, Auskleidung f
covering cap   Abdeckklappe f
covering panel   Verkleidungsblech n
covering plate   Abdeckplatte f
covering sheet   Abdeckblech n
chain covering   Kettenverschalung f
felt covering   Filzauflage f

cowl Stirnwand f, Motorhaube f
  cowl panel Windlauf m
crank/to anlassen
crank Kurbel f, Kröpfung f
  crank arm Kurbelarm m
  crank drive Kurbelantrieb m
  crank gear Kurbelgetriebe n, Kurbeltrieb m
  crank pin Kurbelzapfen m
  driving crank Antriebskurbel f, Schaltkurbel f
  hand crank Handkurbel f
  starting crank Andrehkurbel f
  window crank Fensterkurbel f
crankcase Kurbelgehäuse n, Kurbelkasten m
  crankcase compression Vorverdichtung f
  crankcase front end cover Steuergehäusedeckel m
  crankcase sump Ölwanne f
  crankcase ventilation Kurbelgehäuseentlüftung f
crankshaft Kurbelwelle f
  crankshaft bearing Kurbelwellenlager n
  crankshaft gear Kurbelwellenrad n
  crankshaft journal Kurbelwellenzapfen m
  crankshaft seal Kurbelwellendichtung f
crash Aufprall m
  crash sensor Aufprallsensor m
  crash test Aufprallprüfung f, Aufprallversuch m
  head-on crash Frontalzusammenstoß m
crashfree driving unfallfreies Fahren n
creep Schlupf m, Radschlupf m
creeper speed Kriechgang m
crimp pliers Crimpzange f
cross Quer-
  cross beam Querträger m, Riegel m
  cross-country car Geländewagen m
  cross-country properties Geländegängigkeit f
  cross-draught carburetor Flachstromvergaser m
  cross ply tire Diagonalreifen m
  cross steering rod Querlenkstange f
  tubular cross member Rohrquerträger m
crown Kranz m, Krone f, Lagerdeckel m

crown wheel großes Differentialantriebskegelrad n, konischer Zahnkranz m, Tellerrad n
  spur gear crown Stirnradkranz m
cruise/to mit konstanter Geschwindigkeit fahren
cubic capacity Hubraum m, Hubvolumen n
cuff Manschette f
cup Napf m, Pfanne f, Becher m
  ball cup Kugelpfanne f
  overflow cup Überlaufbecher m
current Strom m
  current collector Stromabnehmer m
  current consumption Stromaufnahme f, Stromverbrauch m
  current delivery Stromabgabe f
  current input Stromaufnahme f
  current intensity Stromstärke f
  current source Stromquelle f
  current strength Stromstärke f
  current supply Stromabgabe f
  alternating current Wechselstrom m
  alternating current generator Wechselstromlichtmaschine f
  battery current Batteriestrom m
  charging current Ladestrom m
  direct current Gleichstrom m
  direct-current generator Gleichstromlichtmaschine f
  discharge current Entladestrom m
  heating current Heizstrom m
  main current switch Hauptstromschalter m
  short-circuit current Kurzschlußstrom m
  starting current Anlaßstrom m
  three-phase current Drehstrom m
cushion Kissen n, Polster n
  first aid cushion Erste-Hilfe-Kissen n
  seat cushion Sitzpolster n
cut off/to abschalten
cut out/to ausschalten
cut-out Schalter m, Ausschalter m
cycle Kreis m, Kreislauf m, Takt m, Prozess m
  cycle of operations Arbeitsspiel n *(beim Motor)*
  ignition cycle Zündzyklus m

cyclone filter   Zyklonfilter n,m
cylinder   Zylinder m
  cylinder arrangement   Zylinderanordnung f
  cylinder block   Zylinderblock m
  compressed air cylinder   Druckluftzylinder m
  control cylinder   Schaltzylinder m, Steuerzylinder m
  door lock cylinder   Türschließzylinder m
  hand brake cylinder   Handbremszylinder m
  lock cylinder   Schloßzylinder m
  main brake cylinder   Hauptbremszylinder m
  oil pressure cylinder   Öldruckzylinder m
  operating cylinder   Betätigungszylinder m
  power cylinder   Arbeitszylinder m *(Servolenkung)*
  single-chamber brake cylinder   Einkammerbremszylinder m
  step cylinder   Stufenzylinder m
  taking cylinder   Nehmerzylinder m
  wheel brake cylinder   Radbremszylinder m
  working cylinder   Arbeitszylinder m *(Servolenkung)*
damp lining   Dämpfungsbelag m
damper   Dämpfer m
  exhaust flame damper   Auspuffflammendämpfer m
  mulileaf damper   Jalousieklappe f
  piston damper   Stoßdämpferkolben m
  plunger damper   Stoßdämpferkolben m
  rubber damper   Gummidämpfer m
  steering damper   Lenkungsdämpfer m
  strut damper   Dämpferbein n
  suspension damper   Stoßdämpfer m, Schwingungsdämpfer m
  tube damper   Stoßdämpferrohr n
  vibration damper   Schwingungsdämpfer m
damping   Dämpfung f
  damping block   Dämpfungsblock m
  damping buffer   Dämpfungspuffer m
  damping plate   Dämpferplatte n
  damping rubber   Dämpfungsgummi m, Stoßdämpfergummi m
  damping sleeve   Dämpfungshülse f
  damping tube   Dämpferrohr n
dashboard   Armaturenbrett n, Instrumentenbrett n,
  Instrumententafel f

dashboard gear change   Armaturenbrettschaltung f, vorgesetzte Schaltung f
dashboard lamp   Armaturen[brett]leuchte f
dashpot   Schwingungsdämpfer m, Stoßdämpfer m; Schließdämpfer m *(beim Vergaser)*
  dashpot carburetor   Vergaser m mit Beschleunigerpumpe
  dashpot pump   Beschleunigerpumpe f *(Vergaser)*
  oil dashpot   ölgefüllter Stoßdämpfer m, Ölbremszylinder m
dazzle   Blendung f
  anti-dazzle   blendfrei
  anti-dazzle device   Blendschutz m
  anti-dazzle mirror   Abblendspiegel m *(elektrisch)*
dead   tot, Tot-
  dead center   Totpunkt m
  dead-center ignition   Totpunktzündung f
  dead point indicator   Totpunktzeiger m
  dead travel   Leerweg m
  dead weight   Eigenmasse f, Leergewicht n
  top dead center   oberer Totpunkt m
deadening   Dämpfung f *(akustisch)*
decelerate/to   Gas wegnehmen, bremsen *(mit Motor)*
deceleration   Geschwindigkeitsverminderung f, Bremsverzögerung f
deck lid   Kofferraumdeckel m
deep bowl reflector   Tiefstrahler m
defective   defekt
deflector   Nase f, Kolbennase f
  deflector plate   Spritzblech n, Ablenkblech n
  air deflector   Ausstellfenster n
  beam deflector switch   Abblendschalter m
  oil deflector   Ölschleuderring m
  oil deflector sheet   Ölabstreifblech n
  telescope deflector   Teleskopaussteller m
defog/to   Beschlag m entfernen
defrost/to   enteisen, entfrosten
defroster   Defroster m, Enteiser m
  defroster hose   Defrosterschlauch m
  defroster system   Enteisungsanlage f
  defroster vent   Entfrosterdüse f
degradation   Zersetzung f, Schädigung f

catalyst degradation Katalysatorschädigung f
degree of saturation Sättigungsgrad m
delay Verzug m, Verzögerung f
  delay switch Verzögerungsschalter m
  angle of delay Zündverzugswinkel m
delayed firing Spätzündung f, Nachzündung f
delivered power abgegebene Leistung f
delivery Ablieferung f, Ausschüttung f
  delivery pipe Einspritzleitung f
  delivery plunger Förderkolben m
  delivery pump Druckpumpe f
  delivery valve Druckventil n
  current delivery Stromabgabe f
demist/to Beschlag m entfernen
demisting Verhindern n des Beschlagens *(Windschutzscheibe)*
demounting device Demontagevorrichtung f
density Dichte f
  acid density Säuredichte f
departure Abfahrt f
  angle of departure hinterer Überhang m
deposit Ablagerung f *(z.B. Ruß)*
desiccator Trockner m, Abgastrockner m
design Design n, Ausführung f, Bauart f, Konstruktion f
  design feature Konstruktionsmerkmal n
  monocoque body design selbsttragende Karosseriebauweise f
desired value Sollwert m
detent Arretierung f, Auslöser m, Synchronkörper m
deterioration Zerstörung f; Zersetzung f; Verschlechterung f,
  Minderung f; Verschleiß m
  catalyst deterioriation Katalysatorzerstörung f
detonate/to knallen; klopfen *(Motor)*
detonation Klopfen n, Klopfgeräusch n
device Vorrichtung f, Gerät n, Einrichtung f
  adjusting device Einstellvorrichtung f
  anti-dazzle device Blendschutz m
  antiglare device Blendschutz m
  centering device Zentriervorrichtung f
  centrifugally operated timing device
  Fliehkraft[zündzeitpunkt]versteller m

centrifugal timing device Fliehkraftversteller m
clamping device Aufspannvorrichtung f, Klemmvorrichtung f,
Spannvorrichtung f
control device Kontrollvorrichtung f
dismounting device Demontagevorrichtung f
ignition device Zündeinrichtung f
injection advance device Einspritzversteller m
mounting device Aufziehvorrichtung f, Montagevorrichtung f
riveting device Nietvorrichtung f
rotary device Drehvorrichtung f
starting device Startanlage f
warning device Warnanlage f
diagnostic Diagnose-, diagnostisch
diagnostic connector Diagnosestecker m
diagnostic plug Diagnosestecker m
diagnostic unit Diagnosegerät n
diagonal tire Diagonalreifen m
diagram Schaubild n, Schema n, Diagramm n
diagram of connections Anschlußschaltbild n
wiring diagram Schaltplan m
dial Zeigerblatt n, Zifferblatt n, Skala f
dial ga[u]ge Meßuhr f
diameter Durchmesser m
internal diameter Innendurchmesser m
outer diameter Außendurchmesser m
diaphragm Membran f
diaphragm cell Membrandose f
diaphragm pump Membranpumpe f
diaphragm spring Membranfeder f, Tellerfeder f
diesel 1.Dieselkraftstoff m; 2.Dieselmotor m
diesel engine Dieselmotor m
diesel oil Dieselkraftstoff m, Dieselöl n
differential Differential n, Ausgleichsgetriebe n; Differential-
differential axle-drive bevel gear Differentialantriebskegelrad n
differential bevel pinion Ausgleichkegelrad n, kleines
Differentialantriebskegelrad n
differential carrier Hinterachsbrücke f, Hinterachsträger m,
Ritzelträger m am Differential
differential case (casing) Ausgleichgehäuse n,

Differentialgehäuse n
differential gear Differential[getriebe] n, Ausgleichgetriebe n
differential housing Differentialgehäuse n
differential lock Ausgleichsperre f, Differentialsperre f
differential master gear [großes] Differentialantriebskegelrad n,
Tellerrad n
differential pinion Antriebskegelrad n, Antriebsritzel n
differential pinion shaft Ausgleichradachse f, Differentialachse f
differential shaft Differentialwelle f
differential spider Ausgleichstern m, Differentialstern m
differential spur gear Ausgleichstirnrad n, Differentialstirnrad n
spur differential Stirnraddifferential n
diffuser Leitapparat m, Ladeleitrad n *(Lader)*
dim/to abblenden
dim light Abblendlicht n
dim switch Abblendschalter m
dimension Abmessung f, Maß n
installation dimension Einbaumaß n
dimmed abgeblendet
dimmed beam Abblendlicht n
dimmer Dimmer m, Abblendschalter m
dimmer pedal Abblendfußschalter m
dimmer switch Abblendschalter m
dimming Abblendung f
dimming cap Abblendkappe f
dimming filament Abblendfaden m
dimming light Abblendlicht n
dimming mirror Abblendspiegel m
dimming switch Abblendschalter m
asymmetric dimming light asymmetrisches Abblendlicht n
foot-operated dimming switch Fußabblendschalter m
headlight dimming Blendschutz m
symmetric dimming light symmetrisches Abblendlicht n
dip/to eintauchen, tauchen
dip Tauch-, Tret-
dip painting Tauchlackierung f
dip switch Abblendschalter m
foot dip switch Fußabblendschalter m
oil dip stick Ölmeßstab m

direct   direkt; Gleich-
  direct and alternating current   Allstrom m
  direct current   Gleichstrom m
  direct current generator   Gleichstromlichtmaschine f
  direct injection   Zylindereinspritzung f, Direkteinspritzung f
direction   Richtung f
  direction indicator   Richtungsanzeiger m,
  Fahrtrichtungsanzeiger m, Winker m
  direction indicator arm   Winkerarm m
  direction-indicator control lamp   Blinkerkontrolleuchte f
  direction of travel   Fahrtrichtung f
directional   gerichtet, ausgerichtet; spurhaltend
  directional control   Spurführung f, Fahrstabilität f
  directional stability   Geradeauslauf m
disc   s. disk
discharge/to   entladen
discharge   Entladung f, Stromentnahme f
  discharge current   Entladestrom m
  discharge jet   Einspritzdüse f
  discharge tube   Spritzrohr n *(am Vergaser)*
  discharge voltage   Entladespannung f
  battery discharge   Batterieentladung f
  glow discharge   Glimmentladung f
  spark discharge   Funkenüberschlag m
discharging characteristic   Entladekennlinie f
disconnect/to   abschalten; trennen
disengage/to   auskuppeln, ausrücken
disengaging   Ausrück-
  disengaging bearing   Ausrücklager n
  disengaging lever   Ausrückhebel m
disk   Scheibe f; Abdeckscheibe f, Bremsscheibe f
  disk brake   Scheibenbremse f
  disk clutch   Scheibenkupplung f, Reibscheibenkupplung f
  disk valve   Tellerventil n, Scheibenventil n
  disk wheel   Scheibenrad n
  adjusting disk   Einstellscheibe f
  clutch disk   Kupplungsscheibe f
  compensating disk   Ausgleichscheibe f
  connecting disk   Anschlußscheibe f

corrugated disk Wellscheibe f
coupling disk Kupplungsscheibe f
fan disk Lüfterscheibe f
friction disk Friktionsscheibe f, Reibscheibe f
hollow disk Hohlscheibe f
spacing disk Distanzscheibe f
stop disk Anschlagscheibe f
synchronizing disk Synchronscheibe f
valve spring disk Ventilfederteller m
valve disk Ventilteller m
dismantle/to demontieren, aus dem Gehäuse herausnehmen
dismount/to demontieren
displacement Verdrängung f, Hubraum m, Hubvolumen n
displacement per cylinder Zylinderhubraum m
power per unit of displacement Hubraumleistung f,
Literleistung f
distance Abstand m, Entfernung f
distance beam headlight Weitstrahler m
distance disk Distanzscheibe f
distance ga[u]ge Distanzlehre f
distance measurement Abstandsmessung f
center distance Achsabstand m
long-distance beam headlight Weitstrahler m
long-distance light Fernlicht n
make-and-break contact distance
Unterbrecherkontaktabstand m
starting distance Anfahrweg m, Anfahrstrecke f
stopping distance Bremsweg m, Bremsstrecke f
distilled water destilliertes Wasser n
distortion Verformung f, Verwindung f
distribution Verteilung f, Verbreitung f
distribution body Verteilerkörper m
distribution box Verteilerdose f
distribution gear Verteilergetriebe n
distribution housing Verteilergehäuse n
distribution of light Lichtverteilung f
distribution shaft Steuerwelle f
distribution tube Verteilerrohr n
distribution valve Steuerventil n

distributor Verteiler m, Zündstromverteiler m, Mengenteiler m *(bei Kraftstoffeinspritzung)*
  distributor arm Verteilerfinger m
  distributor cam Verteilernocken m, Zündverteilernocken m
  distributor cap Verteilerkappe f *(Zündverteiler)*
  distributor case Verteilergehäuse n
  distributor coil Zündspule f
  distributor contact Unterbrecherkontakt m
  distributor disk Verteilerscheibe f
  distributor housing Verteilergehäuse n
  distributor piece Verteilerstück n
  distributor pump Verteilerpumpe f, Verteilereinspritzpumpe f
  distributor rotor Verteilerläufer m, Verteilerfinger m
  distributor rotor arm Verteilerfinger m
  distributor shaft Verteilerwelle f
  distributor valve Verteilerventil n
  ignition distributor Zündverteiler m
  oil distributor tube Ölverteilerrohr n
  plug distributor Steckverteiler m
disturbance Störung f
divided geteilt, unterteilt
  divided combustion chamber unterteilter Brennraum m, Nebenkammerbrennraum m
  divided line brake system Zweikreisbremsanlage f
dog Mitnehmer m, Nase f, Klaue f
  dog clutch (coupling) Klauenkupplung f
  driving dog Antriebsklaue f
door Tür f
  door brake Türbremse f
  door catch Türhalter m, Türanschlag m, Türnase f
  door contact Türkontakt m
  door contact interrupter Türkontaktschalter m
  door frame Türrahmen m
  door guide Türführung f
  door handle Türgriff m
  door hinge Türscharnier n
  door latch Türklinke f, Türriegel m, Türverschluß m
  door lock Türschloß n
  door lock cylinder Türschließzylinder m

door lock latch Drehfalle f
door-operating cylinder Türbetätigungszylinder m
door panel[l]ing Türverschalung f
door sill Türschwelle f
door stiffening Türverstärkung f, Türversteifung f
door stop Türanschlag m
door trim Türverkleidung f
door weather strip Türdichtungsstreifen m
outside door panel Türaußenblech n
rear door Hecktür f
double doppelt, zweifach
double de-clutch zweimaliges Kuppeln n, Zwischenkuppeln n
double disk clutch Zweischeibenkupplung f
double-stage zweistufig
doughnut Gelenkscheibe f, Scheibengelenk n
dowel Dübel m
dowel pin Paßstift m
grooved dowel pin Kerbstift m
downdraught carburetor Fallstromvergaser m
drag Luftwiderstand m, Strömungswiderstand m
drag link Lenkspurhebel m, Lenkstange f
drag rod Lenkschubstange f
aerodynamic drag Luftwiderstand m
coefficient of aerodynamic drag Luftwiderstandsbeiwert m
road drag Rollwiderstand m
drain Abfluß-, Ablaß -
drain cock Ablaßhahn m
drain plug Ablaßschraube f
drain valve Entwässerungsventil n
oil drain pipe Ölablaßstutzen m
draught Strömung, Zug m
cross-draught carburetor Flachstromvergaser m
horizontal draught carburetor Flachstromvergaser m
draw/to ziehen, zeichnen
draw Zug m
draw band Zugband n
draw rod Zugstange f
draw spring Rückzugfeder f, Zugfeder f
drip mo[u]lding Regenrinne f, Regenleiste f, Dachrinne f

drive/to fahren, steuern, lenken
drive Antrieb m, Trieb m
  drive assembly Getriebe n
  drive bearing Antriebslager n
  drive belt Antriebsriemen m
  drive chain Antriebskette f
  drive cover Antriebsdeckel m
  drive flange Antriebsflansch m
  drive housing Schaltgehäuse n
  drive line s. drive shaft
  drive shaft Gelenkwelle f, Kardanwelle f, Antriebswelle f
  all-wheel drive Allradantrieb m
  belt drive Riemenantrieb m
  differential axle-drive bevel gear Differentialantriebskegelrad n
  differential drive Differentialantrieb m
  final drive Hinterradantrieb m
  four-wheel drive Vierradantrieb m, Geländeantrieb m
  front drive Vorderradantrieb m, Frontantrieb m
  front wheel drive Vorderradantrieb m
  gear drive Zahnradantrieb m
  rear drive Heckantrieb m, Hinterradantrieb m
  rear wheel drive Hinterradantrieb m
driven pulley Antriebsscheibe f
driver Fahrer m, Chauffeur m
driver's licence Führerschein m, Fahrerlaubnis f
driving Antriebs-, Fahr-
  driving axle Antriebsachse f
  driving belt Antriebsriemen m
  driving bush Mitnehmerbüchse f
  driving comfort Fahrkomfort m
  driving dog Antriebsklaue f, Mitnehmer m
  driving flange Mitnehmerflansch m
  driving fork Mitnehmergabel f
  driving gear Antriebsrad n, Antriebsklaue f, Mitnehmer m
  driving group Antriebsgruppe f
  driving lamp Fernscheinwerfer m, Weitstrahler m
  driving licence Fahrerlaubnis f, Führerschein m
  driving light Fernlicht n
  driving mirror Rückspiegel m

driving **pawl**   Mitnehmerklinke f
driving **performance**   Fahrleistung f
driving **pinion**   Antriebszahnrad n, Schaltritzel n, Antriebsritzel n
driving **shaft**   Antriebswelle f, Triebachse f, Antriebswelle f
driving **side**   Antriebsseite f
driving **speed**   Fahrgeschwindigkeit f
driving **wheel**   Antriebsrad n
drop   Tropfen m, Abfall m, Kröpfung f *(Rahmen)*
drop **arm**   Lenkstockhebel m, Lenkspurhebel m
drop **base rim**   Tiefbettfelge f
drop **center rim**   Tiefbettfelge f
drop **frame**   Niederrahmen m, Tiefbettrahmen m
frame drop   Rahmenkröpfung f
voltage drop   Spannungsabfall m
drum   Trommel f, Bremstrommel f
drum **brake**   Trommelbremse f
drum **testing machine**   Reifenprüfstand m
clutch drum   Kupplungstrommel f
dry   trocken
dry **multidisk (multiplate) clutch**
Mehrscheibentrockenkupplung f
dry **single-disk (single-plate) clutch**
Einscheibentrockenkupplung f
dr-y **sump**   Trockensumpf m *(Motorschmierung)*
dry **sump lubrication**   Trockensumpfschmierung f
dry **sump system**   Trockensumpfschmieranlage f
dry-plate clutch   Trocken[scheiben]kupplung f
dual   dual, Zwei-, Doppel-
dual **circuit brake system**   Zweikreisbremsanlage f
dual **exhaust**   Doppelauspuff m
dual **ignition**   Zweifachzündung f
dual **stage carburetor**   Stufenvergaser m
dual **tire**   Doppelreifen m, Zwil[l]ingsreifen m
duct   Leitung f, Kanal m
cable duct   Kabelkanal m
durability   Lebensdauer f, Haltbarkeit f
duration   Dauer f
duration **of combustion**   Verbrennungsdauer f
closing duration of contact   Kontaktschließungsdauer f

    load duration Belastungsdauer f
dust Staub m
  dust cap Staubkappe f
  dust envelope Staubhülle f, Staubschutzhülle f
  dust protection Staubschutz m
  dust shield Staubblech n, Staubschutzblech n
dwell Öffnungsdauer f
  dwell angle Schließwinkel m *(Unterbrecher)*
  dwell of cam Nockenerhebung f
dynamo Dynamo m, Lichtmaschine f
  dynamo charging lamp Ladekontrolleuchte f
  alternating-current dynamo Wechselstromdynamo m
early ignition Frühzündung f, Vorzündung f
earth/to erden
earth Masse f, Erdung f *(s.a. ground)*
  earth connection Masseanschluß m
  earth electrode Masseelektrode f *(Zündkerze)*
  earth lead Masseleitung f
  earth strap Masseband n
  earth terminal Masseklemme f
  earth wire Masseleitung f
earthing lead Massekabel n
ease Leichtgängigkeit f *(Lenkung)*
eccentric 1. Exzenter m; 2. exzentrisch, außermittig
eddy current Wirbelstrom m
  eddy current brake Wirbelstrombremse f
edge Kante f
effect Leistung f, Wirkung f, Ergebnis n
  castor effect Nachlauf m
  luminous effect Lichtwirkung f
efficiency Wirkungsgrad m
  light efficiency Lichtausbeute f
  luminous efficiency Lichtausbeute f, Lichtleistung f
efficient leistungsfähig
elbow Rohrkrümmer m, Kniestück n
  connecting elbow Anschlußkrümmer m
  exhaust elbow Abgaskrümmer m, Auspuffkrümmer m
  suction elbow Ansaugkrümmer m
  tube elbow Rohrkrümmer m

electric elektrisch, Elektro-
  electric arc welding Elektroschweißung f, Elektroschweißen n,
  Lichtbogenschweißung f, Lichtbogenschweißen n, elektrisches
  Schweißen n
  electric backlight defroster elektrische Heckscheibenheizung f,
  heizbare Heckscheibe f
  electric conductance Leitwert m
  electric control elektrische Steuerung f
  electric drive Elektroantrieb m
  electric fuel pump elektrische Kraftstoffpumpe f,
  Elektrokraftstoffpumpe f
  electric heated window elektrisch beheizte Heckscheibe f
  electric power supply Stromversorgung f
  electric rear window defroster elektrische Heckscheibenheizung f
  electric rear window demister elektrische
  Heckscheibenheizung f, heizbare Heckscheibe f
  electric windows elektrischer Fensterheber m, elektrischer
  Scheibenheber m
electrode Elektrode f
  electrode burning Elektrodenabbrand m
  central electrode of spark plug Kerzenstift m
  plug electrode Zündkerzenelektrode f
electrolyte Elektrolyt m
electromagnet Elektromagnet m
electromagnetic elektromagnetisch
  electromagnetic cold-start valve elektromagnetisches
  Kaltstartventil n, elektromagnetisches Startventil n,
  Elektrostartventil n
  electromagnetic starter valve elektromagnetisches Startventil n,
  Elektrostartventil n, elektromagnetisches Kaltstartventil n
  electromagnetic switch elektromagnetischer Schalter m
element Element n, Bauteil n; Einsatz m, Patrone f *(Filter)*
  control element Schaltelement n, Steuerelement n
  filter element Filterelement n
  heating element Heizelement n
emergency Notfall m, Not-
  emergency brake Handbremse f, Feststellbremse f
  emergency flasher switch Pannenwarnblinkschalter m,
  Warnblinkschalter m

emergency fuel tank   Kraftstoff-Reserve-Behälter m, Kraftstoff-Reservetank m

emission   Emission f, Ausstrahlung f, Schadstoffausstoß m, Abgas n, Auspuffgas n

emission control   Abgasreinigung f, Abgasentgiftung f

emission limit   Schadstoffgrenzwert m

emission of particulates   Feststoffemission f, Teilchenemission f, Partikelemission f

emulsion tube   Mischrohr n *(am Vergaser)*

end   Ende n, Schluß m, Endpunkt m

end cap   Abdeckkappe f

drive end bearing   Antriebslager n

motorcycle handle-bar end   Kraftradlenkerende n

shaft end   Wellenstumpf m

endurance run   Dauerlauf m *(Prüfung)*

energy   1. Energie f; 2. Arbeitsvermögen n

energy absorbing   energieaufnehmend

aerodynamic energy   Strömungsenergie f

engage/to   eingreifen, einrasten

engage the clutch/to   einkuppeln

engagement   Eingriff m

engaging of the pinion   Ritzeleingriff m, Ritzeleinspurung f

engine   Motor m

engine bay   Motorraum m

engine block   Motorblock m

engine bonnet   Motorhaube f

engine brake   Motorbremse f

engine breakdown   Motorschaden m, Motorausfall m, Motorpanne f

engine capacity   Hubraum m

engine characteristics   Motorcharakteristik f

engine compartment   Motorraum m

engine compartment rear bulk   Stirnwand f, Spritzwand f, Motorraumschott n

engine coolant   Motorkühlmittel n

engine cooling fan   Lüfter m, Ventilator m

engine cowl   Motorhaube f

engine cutout   Motoraussetzer m, Zündaussetzer m, Aussetzer m

engine data   Motordaten pl, Motorkenndaten pl

engine diagnostic plug Motor-Diagnosestecker m
engine displacement Hubraum m
engine dynamometer Motorprüfstand m, Motorenprüfstand m
engine efficiency Motorwirkungsgrad m
engine failure Motorschaden m, Motordefekt m, Motorpanne f
engine firewall Motortrennwand f, Motorraumschott n
engine hood Motorhaube f, Kühlerhaube f
engine hood release Motorhaubenentriegelung f
engine idling Motorleerlauf m
engine indicator Druck-Indiziergerät,
Zylinderdruck-Indiziergerät n, Indiziergerät n
engine installation Motoreinbau m
engine installation angle Motoreinbauwinkel m, Einbauwinkel m
des Motors
engine knock Motorklopfen n
engine knocking Motorklopfen n
engine mount transverse Motorquerträger m
engine mounting Motorlager n, Motorträger m,
Motoraufhängung f
engine number Motornummer f
engine oil cooler Motorölkühler m
engine oil gallery Ölverteilerleitung f, Hauptölleitung f,
Hauptölkanal f
engine oil pressure ga[u]ge Motoröldruckmeßgerät n
engine oil pressure indicator Motoröldruckanzeigegerät n
engine output Motorleistung f
engine power Motorleistung f
engine response Ansprechen n des Motors, Ansprechverhalten n
des Motors
engine specifications Motordaten pl
engine starter Anlasser m
engine stop control Betätigungseinrichtung f zum Abstellen des
Motors, Motorabstellhebel m
engine support bracket Motorlager n
engine suspension Motoraufhängung f, Motorlagerung f
engine test-bed Motorprüfstand m, Motorenprüfstand m
engine timing Motorsteuerung f, Ventiltrieb m, Ventilantrieb m
engine torque Motordrehmoment n
driving engine Antriebsmaschine f

horizontally opposed-piston engine   Boxermotor m, Gegenkolbenmotor m

internal combustion engine   Verbrennungsmotor m

rear engine   Heckmotor m

supercharged engine   aufgeladener Motor m

engineering   Technik f

automotive engineering   Kraftfahrzeugtechnik f

enrich/to   anreichern *(Gemisch)*

enrichment   Anreicherung f

enrichment device   Anreicherungssystem n

enrichment jet   Anreicherungsdüse f

entrance   Einstieg m

entrance lamp   Einstiegsleuchte f, Türleuchte f

envelope   Hülle f, Ummantelung f

dust envelope   Staubhülle f, Staubschutzhülle f

protecting envelope   Schutzhülle f

equipment   Ausrüstung f, Ausstattung f

charging equipment   Ladeanlage f

interference suppression equipment   Entstörsatz m

signal[l]ing equipment   Signalanlage f

test equipment   Prüfeinrichtung f

windscreen wiper equipment   Scheibenwischeranlage f, Scheibenwaschanlage f

estate car   Kombifahrzeug n, Kombiwagen m

evaporate/to   verdunsten, verdampfen

evaporator   Verdampfer m

excess   Überschuß m, Mehr-, Über-

excess air   Luftüberschuß m, Überschußluft f

excess weight   Mehrgewicht n, Übergewicht n

excessive pressure   Überdruck m

exchangeable   auswechselbar

exchange unit   Austauschteil n

exchanger   Austauscher m, Übertrager m

heat exchanger   Wärmeübertrager m, Wärmeaustauscher m

excite/to   erregen

exciting   Erreger-

exciting coil   Erregerspule f

exciting current   Erregerstrom m

exhaust   Auspuff m

exhaust air   Abluft f
exhaust and refill   Ladungswechsel m, Gaswechsel m
exhaust bend   Austrittskrümmer m
exhaust casing   Abgasgehäuse n
exhaust cycle   Auspufftakt m
exhaust-driven supercharger   Abgasturbolader m, Turbolader m
exhaust duct   Auslaßkanal m
exhaust elbow   Abgaskrümmer m, Auspuffkrümmer m
exhaust emission   Abgas n, Auspuffgas n
exhaust emission control   Abgasreinigung f, Abgasentgiftung f
exhaust emission control system   Abgasreinigungsanlage f,
Abgasentgiftungsanlage f
exhaust engine brake   Auspuffmotorbremse f
exhaust flange   Auspuffflansch m
exhaust gas   Abgas n, Auspuffgas n,
exhaust gas catalyzer   Abgaskatalysator m
exhaust gas limit   Abgasgrenzwert m
exhaust gas probe   Abgassonde f
exhaust gas purifier   Abgasreiniger m
exhaust gas recirculation   Abgasrückführung f
exhaust gas system   Abgasanlage f
exhaust gas turbocharger   Abgasturbolader m, Turbolader m
exhaust heat   Abgaswärme f
exhaust-jacket carburetor   Vergaser m mit Auspuffheizmantel
exhaust manifold   Auspuffkrümmer m
exhaust muffler   Abgas-Schalldämpfer m,
Auspuff-Schalldämpfer m, Auspufftopf m
exhaust-operated air heating   Abgasheizung f
exhaust pipe   Auspuffleitung f, Auspuffrohr n, Abgasleitung f,
Abgasrohr n
exhaust pipe clamp   Auspuffrohrschelle f
exhaust piping   Auspuffleitung f
exhaust port   Auslaßkanal m, Auslaßsteueröffnung f
(Kreiskolbenmotor)
exhaust port liner   Auslaßkanalbuchse f
exhaust side   Auslaßseite f
exhaust silencer   Auspuffschalldämpfer m, Auspufftopf m
exhaust smoke   Abgasrauch m
exhaust stroke   Auspuffhub m, Auspufftakt m, Auslaßhub m

exhaust system Auspuffanlage f, Abgasanlage f
exhaust tube Auspuffleitung f, Auspuffrohr n, Abgasleitung f,
Abgasrohr n
exhaust turbocharger Abgasturbolader m, Turbolader m
exhaust valve Auslaßventil n, Auspuffventil n
exhaust valve cap Auslaßventilaufsatz m,
Auslaßventilverschraubung f
exit Ausstieg m, Ausgang m
light exit Lichtaustritt m
expander Andrückfeder f
expansion Ausdehnung f
expansion container (reservoir) Expansionsgefäß n,
Expansionsbehälter m, Ausgleichsbehälter m
pre-expansion chamber Auspuffvorschalldämpfer m
thermal expansion Wärmeausdehnung f
extension Verlängerung f
extension device Ausziehvorrichtung f
frame extension Rahmenverlängerung f
external ventilation Außenbelüftung f *(am Vergaser)*
extra headlamp Zusatzscheinwerfer m
extras Sonderzubehör n, Extras npl
fabric Gewebe n, Stoff m
fabric break Gewebebruch m
fabric layer (ply) Gewebelage f *(Reifen)*
face Stirn[seite] f , Vorderseite f
joint face Dichtfläche f *(Motorgehäuse)*
slide face Gleitfläche f
facia Armaturenbrett n, Instrumententafel f
fade Bremsschwund m
fading Schwund m; Bremsschwund m
fading control Überblendregelung f *(Radio)*
Kupplungsschwund m
failure Ausfall m, Defekt m
engine failure Motordefekt m
false start Fehlstart m
fan Lüfter m, Ventilator m, Gebläse n
fan belt Ventilatorriemen m
fan blade Ventilatorflügel m
fan clutch Lüfterkupplung f

fan cowl Kühlerzarge f, Lüfterzarge f
fan guard Lüfterzarge f
fan motor Lüftermotor m
fan pulley Ventilatorscheibe f
fan shaft Ventilatorwelle f, Lüfterwelle f
fan shroud Lüfterzarge f
fan wheel Lüfterrad n
fast back tail Fließheck n, Abreißheck n
fasten/to befestigen
fastening Befestigung f
  fastening bolt Befestigungsbolzen m
  fastening screw Befestigungsschraube f
feather key Federkeil m
feed Zuführungs-
  feed line Zuleitung f, Zulaufleitung f
feeler Fühler m, Taster m
  feeler ga[u]ge Fühlerlehre f, Abstandslehre f
  temperature feeler Temperaturfühler m
felt Filz m
  felt gasket Filzdichtung f
  felt joint Filzdichtung f
  felt washer Filzunterlegscheibe f
female socket Steckbuchse f
fender Kotblech n, Schutzblech n, Kotflügel m
  fender flave Kotflügelverbreiterung f
  front fender Stoßstange f
fifth fünfter, fünfte, fünftes
  fifth wheel Schlepprad n, Sattelkupplung f, Aufsattelkupplung f
  fifth-wheel load Aufsattellast f, Aufsatteldruck m
  fifth-wheel steering Drehschemellenkung f
filament Glühdraht m, Glühfaden m
  dimming filament Abblendfaden m
filler Füller m, Füllvorrichtung f, Füllstoff m
  filler neck Einfüllstutzen m
  filler plug Einfüllschraube f, Füllschraube f
  oil filler neck Öleinfüllstutzen m
  radiator filler cap Kühlerdeckel m
filling capacity Füllmenge f
filling screw Einfüllschraube f *(z.B. für Öl)*

film   Film m, Überzug m
  lubricating film   Schmierfilm m
  oil film   Schmierfilm m
filter   Filter n,m
  filter bowl   Filtertopf m, Filtergehäuse n
  filter cartridge   Filterpatrone f, Filtereinsatz m
  filter element   Filterelement n
  filter gauze   Filtergaze f
  filter unit   Filtereinsatz m
  air filter   Luftfilter n,m
  air filter intake silencer   Luftfilteransaugschalldämpfer m
  change filter   Wechselfilter n,m
  compressed air filter   Druckluftreiniger m
  fuel filter   Brennstoffilter n,m Kraftstoffilter n,m
  full-flow filter   Hauptstromfilter n,m
  induction air filter   Ansaugluftfilter n,m
  oil filter   Ölfilter n,m
  partial-flow filter   Nebenstromfilter n,m
  preliminary filter   Vorfilter n,m
fin   Rippe f, Leitflügel m
final   End-, Abschluß -
  final assembly   Endmontage f
  final coat   Decklack m
  final cover   Abschlußdeckel m
  final cut-out   Endausschalter m
  final drive   Hinterradantrieb m
  final-reduction gear   Radvorgelege n
fine filter   Feinfilter n,m
  fine screen filter   Feinfilter n,m
finger   Finger m
  finger-tip control   Betätigung f durch Fingerspitze,
  Fingerdruckbetätigung f
  release finger   Auslösefinger m
finish   Ende n, Oberflächenbeschaffenheit f, Deckenanstrich m
  surface finish   Oberflächenausführung f
  tropical finish   Tropenausführung f
finned   lamelliert, gerippt
  finned radiator   Lamellenkühler m, Rippenkühler m
  finned tube   Lamellenrohr n, Rippenrohr n

fire/to zünden
fire extinguisher Feuerlöscher m
fire-extinguishing truck Löschfahrzeug n
firewall Stirnwand f, Spritzwand f
firing Zündung f
  firing order (sequence) Zündfolge f
  firing point Zündzeitpunkt m
first Erst-
  first-aid cushion Erste-Hilfe-Kissen n
  first-aid kit Erste-Hilfe-Ausrüstung f, Verbandskasten m
fit Passung f, Sitz m
  thread fit Gewindepassung f
fitting Montage f, Anbringung f, Anschlußstück n, Nippel m,
  Beschlag m
  fitting bolt Anschlußbolzen m
  connection fitting Anschlußnippel m
  tire fitting Reifenmontage f, Reifenreparatur f
  wide-angle lighting fitting Breitstrahler m
fixed caliper Festsattel m *(Bremse)*
flange Flansch m, Felgenhorn n
  connecting flange Anschlußflansch m
  drive flange Antriebsflansch m
  driving flange Mitnehmerflansch m
  gear box flange Getriebeflansch m
flap Klappe f
  air flap Luftklappe f
  covering flap Deckklappe f
  rear flap Heckklappe f
flare/to flackern, sich bauschen, verbreitern
flared fenders verbreiterte Kotflügel mpl
flash/to blinken
flash Blitz m, Aufblitzen n
  flash control Blinkkontrolle f
  flash signal Lichthupe f, Blinkzeichen n
  flash-tail lamp Blink-Schlußleuchte f
flasher Blinker m, Blinkgeber m, Lichthupe f
  flasher indicator switch Blinkschalter m
  flasher lamp Blinkleuchte f, Blinker m
  flasher unit Blinkgeber m

flasher warning light   Blinkkontrollampe f
  rear flasher lamp   Heckblinkleuchte f, Schlußblinkleuchte f
  side-flasher lamp   Begrenzungsblinkleuchte f
  tail flasher lamp   Schluß-Blinkleuchte f
flat   1.flach; 2.Pannenreifen m, "Platter" m
  flat-base rim   Flachbettfelge f
  flat belt   Flachriemen m
  flat engine   Boxermotor m
  flat piston crown (head)   flacher Kolbenboden m
  flat plug   Flachstecker m
  flat plug box   Flachsteckdose f
  flat round screw   Flachrundschraube f
  flat spot   Beschleunigungsloch n, "Loch" n beim Gasgeben
  flat spotting   Abplattung f *(Reifen)*
  flat spring   Blattfeder f, Flachfeder f
  flat tire radius   Halbmesser m des platten Reifens
  flat-top piston   Kolben m mit flachem Boden, Flachkolben m
  flat tubular connector   Flachsteckverbinder m
  flat tubular radiator   Flachrohrkühler m
flexible   flexibel, beweglich; elastisch
  flexible coupling   elastische Kupplung f
  flexible metal tube   Metallschlauch m
  flexible pipe   biegsames Rohr n
  flexible shaft   biegsame Welle f
flexing   Walken n *(Reifen)*
  flexing area   Walkzone f
  flexing resistance   Walkwiderstand m
flick wiper   Intervallscheibenwischer m
float   Schwimmer m *(Vergaser)*
  float bowl   Schwimmerkammer f, Schwimmergehäuse n
  float chamber   Schwimmergehäuse n
  float chamber valve   Schwimmernadelventil n
  float housing   Schwimmergehäuse n, Schwimmerkammergehäuse n
  float needle   Schwimmernadel f
  float-needle valve   Schwimmernadelventil n
floating   Schwebe-, Schwing-
  floating axle   Schwingachse f, Schwebeachse f
  floating caliper   Schwimmsattel m, Faustsattel m *(an der Scheibenbremse)*

floating caliper disk brake Schwimmsattel-Scheibenbremse f, Faustsattel-Scheibenbremse f
floor Boden m
  floor carpet Bodenteppich m, Bodenbelag m
  floor covering Fußraumauskleidung f, Bodenbelag m
  floor mat Bodenmatte f, Fußbodenmatte f
  floor pan (panel) Bodenplatte f, Bodenblech n
  floor switch Tretschalter m
flow Strömung f, Durchflußmenge f
  flow indicator Strömungsanzeiger m, Durchsatzanzeiger m
  flow meter Durchflußmesser m
  flow rate Durchflußrate f, Durchsatz m
  full-flow filter Hauptstromfilter n,m
  partial-flow filter Nebenstromfilter n,m
fluid Flüssigkeit f, Fluid n *(Gas oder Flüssigkeit)*
  fluid coupling Flüssigkeitskupplung f, hydraulische Kupplung f
  fluid level Flüssigkeitsstand m
  fluid reservoir Ausgleichbehälter m *(z.B. für Bremsflüssigkeit)*
flyweight Schwungmasse f
flywheel Schwungrad n, Schwungscheibe f
  flywheel gear ring Schwungradzahnkranz m
  flywheel hub Schwungradnabe f
  flywheel rim Schwungradkranz m
  flywheel starter Schwungradanlasser m
  flywheel timing mark Schwungradmarke f
foam Schaum m
  foam padding Umschäumung f
  foam rubber Schaumgummi m
fog Nebel m
  fog lamp Nebelscheinwerfer m
  fog tail light Nebelschlußleuchte f
foil Folie f
fold-down rear bench umklappbare Rücksitzbank f
folding Falt-, Klapp-
  folding bow Faltspriegel m
  folding hood Faltdach n, Klappverdeck n
  folding top Faltdach n, Klappverdeck n
follower Stößel m, Nockenstößel m
foot Fuß m

foot board   Trittbrett n, Trittplatte f
foot brake   Fußbremse f, Betriebsbremse f
foot control   Fußhebel m, Pedal n
foot control linkage   Fußhebelwerk n
foot dip switch   Fußabblendschalter m
foot lever   Fußhebel m
foot-operated dimming switch   Fußabblendschalter m
foot-operated switch   Fußschalter m
foot rest   Fußstütze f, Fußraste f
foot switch   Fußschalter m
force   Kraft f
  aerodynamic force   Luftwiderstandskraft f
  cornering force   Seitenkraft f
  lateral force   Seitenführungskraft f, Spurhaltungskraft f
  tensile force   Zugkraft f
fork   Gabel f
  gear-shift fork   Schaltgabel f
forward   Vorwärts-, Front-
  forward control vehicle   Frontlenker m
  forward gear   Vorwärtsgang m
  forward speed   Vorwärtsgang m
four   vier
  four barrel carburetor   Vierfachvergaser m,
  Doppelregistervergaser m
  four-cycle engine   Viertaktmotor m
  four-speed gearbox (transmission)   Vierganggetriebe n
  four-stroke engine   Viertaktmotor m
  four-wheel drive   Vierradantrieb m, Geländeantrieb m
frame   Rahmen m
  frame construction   Rahmenbauweise f
  frame extension   Rahmenverlängerung f
  frame reinforcement   Rahmenverstärkung f
  frame side bar   Längsträger m, Rahmenholm m
  battery frame   Batterierahmen m
  central tube frame   Mittelrohrrahmen m
  chassis frame   Fahrgestellrahmen m, Karosserierahmen m
  coil frame   Spulenhalter m
  intermediate frame   Zwischenrahmen m
  roof frame   Dachrahmen m

seat frame   Sitzgestell n
supporting frame   Stützrahmen m
tubular frame   Rohrrahmen m
frameless   rahmenlos, selbsttragend
   frameless body   selbsttragende Karosserie f
   frameless construction   selbsttragende Konstruktion f, rahmenlose Bauart f
framework   Gerippe n *(Karosserie)*
free   frei
   free motion   Spiel n
   free travel   Kupplungsspiel n
   free wheeling   Freilauf m
freeze/to   gefrieren, erstarren
   anti-freeze   Frostschutz m
fresh   frisch
   fresh air duct   Frischluftkanal m
   fresh air screen   Frischluftgitter n
friction   Reibung f
   friction bearing   Gleitlager n
   friction clutch   Rutschkupplung f
   friction lining   Reibbelag m
   friction pad   Bremsbelag m
   multi-plate friction clutch   Lamellenrutschkupplung f
   road friction   Fahrbahnreibung f
front   Front-, Vorder-
   front axle   Vorderachse f
   front brake   Vorderradbremse f
   front drive   Vorderradantrieb m, Frontantrieb m
   front end   Karosserievorbau m
   front fender   vorderer Kotflügel m
   front muffler   Vorschalldämpfer m
   front panel   Vorderwand f
   front passenger   Beifahrer m
   front silencer   Vorschalldämpfer m
   front suspension   Vorderradaufhängung f
   front wall   Vorderwand f, Stirnwand f
   front wheel   Vorderrad n
   front wheel drive   Vorderradantrieb m
   front wing   Vorderkotflügel m, vorderer Kotflügel m

independent front suspension   Einzelradaufhängung f
fuel/to   tanken; betanken
fuel   Kraftstoff m, Brennstoff m, Treibstoff m
  fuel additive   Kraftstoffzusatz m, Additiv n
  fuel-air mixture   Kraftstoff-Luft-Gemisch n
  fuel-air ratio   Luftzahl f
  fuel-air ratio control unit   Gemischregler m
  fuel atomization   Kraftstoffzerstäubung f, Kraftstoffvergasung f
  fuel baffle   Schwallblech n *(im Kraftstoffbehälter)*
  fuel capacity   Tankinhalt m
  fuel carburation   Kraftstoffvergasung f, Kraftstoffzerstäubung f
  fuel cell   Brennstoffzelle f
  fuel comsumption   Kraftstoffverbrauch m
  fuel-consumption ga[u]ge   Kraftstoffverbrauchsmesser m
  fuel cut-off system   Kraftstoffabschaltung f
  fuel delivery   Kraftstoffförderung f
  fuel demand   Kraftstoffverbrauch m
  fuel discharge arm   Kraftstoffaustrittsarm m
  fuel distributor   Mengenteiler m, Kraftstoffdosiereinrichtung f
  fuel economy   Kraftstoffwirtschaftlichkeit f, Kraftstoffsparsamkeit f
  fuel evaporation control valve   Kraftstoffverdunstungsventil n,
Tankbelüftungsventil n, Absaugventil n
  fuel exit point   Kraftstoffaustrittsstelle f
  fuel feed   Kraftstoffzufuhr f, Kraftstoffversorgung f,
Kraftstoffförderung f
  fuel-feed pressure   Kraftstoff-Förderdruck m,
Kraftstoffleitungsdruck m, Druck m in der Kraftstoffleitung
  fuel feed pump   Kraftstoffförderpumpe f, Kraftstoffpumpe f
  fuel filler cap   Tankklappe f
  fuel filler neck   Kraftstoffeinfüllstutzen m
  fuel filler pipe   Kraftstoffeinfüllstutzen m
  fuel filter   Kraftstoffilter n,m
  fuel flow   Kraftstoffdurchsatz m, Kraftstoffdurchfluß m
  fuel flow meter   Kraftstoffdurchsatzmesser m,
Kraftstoffdurchflußmesser m
  fuel flow rate   Kraftstoffdurchsatz m, Kraftstoffdurchfluß m
  fuel ga[u]ge   Vorratsgeber m, Kraftstoffanzeigegerät n,
Kraftstoffvorratsanzeige f
  fuel grade   Kraftstoffsorte f

fuel heating Kraftstoffheizung f
fuel hose Kraftstoffschlauch m
fuel indicator Kraftstoffanzeige f, Kraftstoffuhr f,
Kraftstoffvorratsanzeige f
fuel inhibitor Klopfbremse f
fuel injection Kraftstoffeinspritzung f, Einspritzung f
fuel injection pump Einspritzpumpe f
fuel-injection system Kraftstoff-Einspritzanlage f,
Einspritzanlage f
fuel injector Einspritzventil n
fuel jet Kraftstoffdüse f
fuel level indicator Kraftstoffstandanzeigegerät n,
Kraftstoffvorratsanzeigegerät n
fuel level warning light Kraftstoffreserve-Warnleuchte f
fuel line Kraftstoffleitung f
fuel flow meter Kraftstoffdurchflußmesser m,
Kraftstoffverbrauchsmeßgerät n
fuel metering Kraftstoffzumessung f, Kraftstoffzuteilung f
fuel metering port (slit) Steuerschlitz m
fuel mileage Kraftstoffergiebigkeit f *(auf Meilen bezogen)*
fuel oil Dieselkraftstoff m, Dieselöl n, Dieseltreibstoff m, Diesel n,
Heizöl n
fuel preparation Kraftstoffaufbereitung f
fuel pump Kraftstoffpumpe f
fuel reserve indicator lamp Kraftstoffreserve-Anzeigeleuchte f
fuel reserve warning lamp Kraftstoffreserve-Warnleuchte f
fuel return Kraftstoffrücklauf m
fuel return line Kraftstoffrücklaufleitung f,
Kraftstoffrückführungsleitung f, Leckkraftstoffleitung f, Leckölleitung f
fuel saving Kraftstoffersparnis f, Kraftstoffeinsparung f
fuel sender Kraftstoffvorratsgeber m, Vorratsgeber m,
Geber-Kraftstoffanzeige f
fuel spillage Kraftstoffüberlaufen n, Kraftstoffauslaufen n,
Kraftstoffleck n
fuel strainer Kraftstoffsieb n
fuel supply Kraftstofförderung f
fuel supply at full load Vollastförderung f, Vollastfördermenge f
fuel-supply line Kraftstoffleitung f
fuel supply pressure Förderdruck m, Kraftstoffpumpendruck m

fuel system Kraftstoffanlage f
fuel tank Kraftstoffbehälter m, Kraftstofftank m, Tank m
fuel tank cap Tankdeckel m
fuel tank capacity Tankinhalt m
fuel tank ga[u]ge Vorratsgeber m
fuel tank lock Tankschloß n
fuel tank ventilation Kraftstoffbehälterentlüfung f,
Kraftstoffbehälterbelüftung f, Kraftstofftankentlüftung f,
Kraftstofftankbelüftung f
fuel tank vent line Kraftstoffbehälterentlüftungsleitung f,
Kraftstoffbehälterbelüftungsleitung f,
Kraftstofftankentlüftungsleitung f, Kraftstofftankbelüftungsleitung f
fuel vapo[u]r lock Dampfblasenbildung f in der Kraftstoffleitung
full Voll-
full advance Frühzündung f, Vorzündung f
full beam Fernlicht n
full charge Volladung f
full-flow filter Hauptstromfilter n,m
full load Vollast f
full throttle Vollgas n
full throttle lock Vollgasanschlag m
funnel Trichter m
funnel bowl Trichterschale f
suction funnel Ansaugtrichter m, Saugtrichter m
fuse/to durchbrennen *(Sicherung)*
fuse Sicherung f, Schmelzsicherung f
fuse box Sicherungsdose f
fuse cabinet Sicherungskasten m
fuse plug Schmelzeinsatz m, Sicherungseinsatz m
fuse strip Schmelzstreifen m, Sicherungsstreifen m
fuse wire Sicherungsdraht m
gage s. gauge
gap Spalt m, Öffnung f
air gap Luftspalt m
garbage collector Müllwagen m, Müllabfuhrwagen m
gas 1. Gas n 2. Benzin n *(US) (s.a. gasoline)*
gas-filled spring device Gasdruckfeder f
gas lever Gashebel m
gas pressure shock absorber Gasdruckstoßdämpfer m

gas tank   Benzintank m
gas turbine   Gasturbine f
exhaust gas   Abgas n
exhaust gas catalyzer   Abgaskatalysator m
gasket   Dichtring m, Dichtung f
  cylinder head gasket   Zylinderkopfdichtung f
  exhaust gasket   Auspuffdichtung f
  oil sump gasket   Ölwannendichtung f
  suction gasket   Ansaugdichtung f
gasoline   Benzin n, Ottokraftstoff m, Vergaserkraftstoff m
  gasoline-air mixture   Benzin/Luft-Gemisch n
  gasoline consumption   Benzinverbrauch m
  gasoline-dispensing pump   Benzin-Zapfsäule f, Zapfsäule f
  gasoline engine   Benzinmotor m
  gasoline hose   Benzinschlauch m
  gasoline injection   Benzineinspritzung f
  gasoline pump   Benzin-Zapfsäule f, Zapfsäule f, Benzinpumpe f
gate   Gatter n; Kulisse f, Schaltkulisse f *(Getriebe)*
  gate change   Kulissenschaltung f
  tail gate   Hecktür f, Heckklappe f
gauge   Anzeigegerät n, Lehre f *(auch: gage)*
  air pressure ga[u]ge   Druckluftmanometer n, Luftdruckprüfer m
  (für Reifen)
  dial ga[u]ge   Meßuhr f
  distance ga[u]ge   Distanzlehre f
  feeler ga[u]ge   Fühlerlehre f
  fuel ga[u]ge   Kraftstoffmeßgerät n, Kraftstoffmesser m
  oil pressure ga[u]ge   Öldruckmesser m
  setting ga[u]ge   Einstell-Lehre f
  valve clearance ga[u]ge   Ventilspiellehre f
gauze filter   Siebfilter n,m
gear/to   schalten *(Getriebe)*
  gear up/to   aufwärts schalten *(Getriebe)*
  gear down/to   abwärts schalten *(Getriebe)*
gear   Zahnrad n, Getriebe n, Gang m
  gear box   Getriebegehäuse n, Getriebekasten m, Wechselgehäuse n
  gear box flange   Getriebeflansch m
  gear case   Getriebegehäuse n
  gear-case cover   Getriebegehäusedeckel m

gear casing   Getriebegehäuse n
gear change   Schaltvorgang m, Gangwechsel m
gear-change box   Wechselgetriebe n
gear change lever knob   Schalthebelknopf m
gear change lever position   Schalthebelstellung f
gear change mechanism   Schaltmechanismus m, Schaltung f
gear changing   Gangwechsel m
gear control fork   Schaltgabel f
gear drive   Zahnradantrieb m
gear-driven supercharger   Getriebelader m
gear lever   Schalthebel m, Getriebeschalthebel m
gear lever bracket   Schaltbock m
gear lever knob   Schalthebelknopf m
gear lever lock   Schalthebelsperre f
gear pump   Zahnradpumpe f
gear rack   Zahnstange f
gear ratio   Übersetzungsverhältnis n, Untersetzungsverhältnis n
gear reduction   Untersetzung f
gear ring   Starterzahnkranz m, Zahnkranz m
gear selector mechanism   Gangwähler m
gear shaft   Zahnradwelle f
gear shift   Schaltvorgang m, Gangwechsel m
gear-shift bar   Schaltstange f
gear-shift cover   Schaltdeckel m
gear-shift fork   Schaltgabel f
gear-shift lever   Schalthebel m, Handschalthebel m
gear shift lever bracket   Schaltbock m
gear shift lever knob   Schalthebelknopf m
gear shift lever shaft   Schalthebelwelle f, Schaltwelle f
gear shift linkage   Schaltgestänge n
gear-shift mechanism   Gangschaltung f
gear-shift pattern   Ganganordnung f
gear-shift rod   Schaltstange f
gear-shift sleeve   Schaltmuffe f
gear wheel   Einrückrad n, Zahnrad n, Getrieberad n
ball-and-socket gear change   Kugelschaltung f
ball-and-socket type gear shifting   Kugelschaltung f
bevel gear couple   Ritzel u. Tellerrad n
cam-and-peg steering gear   Schraubenganglenkung f,

Spindellenkgetriebe n
camshaft timing gear   Nockenwellenantriebsrad n
change-speed gear   Wechselgetriebe n
differential gear   Differential[getriebe] n, Ausgleichgetriebe n
differential master gear   *(großes)* Differentialantriebskegelrad n
differential spur gear   Ausgleichstirnrad n, Differentialstirnrad n
distribution gear   Verteilergetriebe n
driving gear   Antriebsrad n, Antriebsklaue f, Mitnehmer m
final gear reduction in the rear axle   Hinterachsübersetzung f
final-reduction gears   Radvorgelege n
flywheel with starting [ring] gear   Schwungrad n mit
Anlasserzahnkranz
forward gear   Vorwärtsgang m
hydraulic gear change   Druckölschaltung f
hydro steering gear   Hydrolenkgetriebe n
intermediate gear   Zwischengetriebe n, Zwischenrad n
master gear   großes Differentialantriebskegelrad n
multi-step reduction gear   Stufengetriebe n
rack-and-pinion steering gear   Zahnstangenlenkgetriebe n
recirculating ball-steering gear   Lenkung f mit Kugelring
reverse gear   Rückwärtsgang m, Rücklaufrad n
running gear   Fahrgestell n, Chassis n, Fahrwerk n
screw-and-nut steering gear   Lenkgetriebe n mit
Schraubenspindel und Mutter, Schraubenlenkung f,
Spindellenkgetriebe n
separate gear change   getrennte Schaltung f
sliding gear   Schieberad n, verschiebbares Zahnrad n
sliding-mesh gear   Schieberad n
speed-change gear   Gangschaltung f, Schaltgetriebe n
starter ring gear   Anlasserzahnkranz m
steering gear case   Lenkgehäuse n
synchromesh gear   Synchrongetriebe n
throw the gear into reverse/to   den Rückwärtsgang einlegen
worm-and-wheel steering gear   Schneckenlenkung f
worm gear   Schneckengetriebe n, Schraubenrad n
gearbox   Schaltgetriebe n, Getriebe n
   gearbox main shaft   Getriebehauptwelle f, Hauptwelle f
   gearbox oil cooler   Getriebeölkühler m
   gearbox shaft   Getriebewelle f

gearbox tunnel Getriebetunnel m, Kardantunnel m, Mitteltunnel m
gearing Übersetzung f
  gearing wheel Übersetzungsrad n
  helical gearing Schrägverzahnung f
generator Dynamomaschine f, Lichtmaschine f
  generator charging indicator Ladekontrollanzeigegerät n
  generator pulley Lichtmaschinenriemenscheibe f
  alternating-current (a.c.) generator Wechselstromlichtmaschine f
  direct-current [d.c.] generator Gleichstromlichtmaschine f
get-away power Anzugsvermögen n
gill Kühlerlamelle f, Kühlrippe f
gilled radiator Lamellenkühler m
glare/to blenden
glass Glas n
  glass fiber reinforced plastic glasfaserverstärkter Kunststoff m
  glass pane Glasscheibe f
  glass rim profile Scheibenfaßprofil n
  glass window Glasscheibe f
  diffusing glass Streuscheibe f
  frosted glass Mattglas n, Milchglas n
  light diffusing glass Lichtstreuscheibe f
  ribbed glass Riffelglas n
  safety glass Sicherheitsglas n
  window glass Fensterglas n
  windscreen glass Windschutzglas n
glazing Verglasung f
glove Handschuh m
  glove box (compartment) Handschuhkasten m, Handschuhfach n
glow/to glühen
glow Glüh-, Heiz-
  glow filament Glühspirale f, Heizwendel f
  glow ignition Glühzündung f
  glow plug Glühkerze f
glue Kleber m, Klebstoff m
governor Regler m
  governor housing Reglergehäuse n
  governor setting Reglereinstellung f

centrifugal governor   Fliehkraftregler m
variable-speed governor   Verstellregler m
gradability   Steigfähigkeit f
grade of accuracy   Güteklasse f
gradient   Steigung f, Neigung f
grease   Fett n
  acid-proof grease   Säureschutzfett n
  lubricating grease   Schmierfett n
grey   grau
  grey cast iron   Grauguß m
  grey cast iron crankcase   Grauguß-Kurbelgehäuse n
grille   Ziergitter n, Kühlergrill m
  radiator grille   Kühlergitter n, Kühlergrill m
grip   Griff m, Handgriff m; Griffigkeit f *(Reifen)*
  grip handle   Haltegriff m
  adjusting grip   Stellgriff m
  ground-grip tire   Geländereifen m
  road grip   Kraftschluß m zur Straße *(Laufstreifen beim Reifen)*
  window grip   Fenstergriff m
grommet   Kabeleintrittsdichtung f, Tülle f, Durchgang m
  cable grommet   Kabeltülle f
  rubber grommet   Gummitülle f
groove   Rille f, Nut f, Profilrille f *(Reifen)*
  key groove   Keilnut f
  oil groove   Schmiernut f, Ölrinne f
grooved dowel pin   Kerbstift m
gross   Brutto-, Gesamt-
  gross vehicle weight   Fahrzeuggesamtgewicht n
  gross weight   zulässiges Gesamtgewicht n
ground/to   erden
ground   Masse f, Erdung f, Boden m *(s.a. earth)*
  ground cable   Massekabel n
  ground clearance   Bodenfreiheit f, Rahmenhöhe f
  ground connection   Masseanschluß m
  ground electrode   Masseelektrode f
  ground-grip tire   Geländereifen m
  ground plate   Grundplatte f
  ground strap   Masseband n
  ground terminal   Masseklemme f

ground wire Masseleitung f
accidental ground Masseschluß m
guard Schutzvorrichtung f
   guard plate Schutzblech n
   splash guard Spritzschutz m
   wheel guard Räderschutz m
gudgeon pin Kolbenbolzen m, Kolbenstift m, Achsbolzen m,
   Kolbenzapfen m
guide Führung f
   guide bolt Führungsbolzen m, Lenkbolzen m
   guide bush Führungsbuchse f, Führungsbüchse f
   guide lever Führungshebel m, Lenkhebel m, Leithebel m
   guide pin Führungszapfen m, Führungsstift m
   guide plate Führungsplatte f
   guide-pulley Leitscheibe f, Leitblock m, Umlenkrolle f
   guide rail Führungsschiene f
   guide rod Führungsstange f
   guide roller Führungsrolle f, Führungswalze f, Leitrolle f,
   Lenkrolle f
   guide tube Führungsrohr n
   cable guide Kabelführung f
   piston guide Kolbenführung f
   sheet-metal guide Leitblech n
   valve guide Ventilführung f
   valve tappet guide Stößelführung f
   window guide Fensterführung f
gulp valve *(s. air gulp valve)*
gutter Regenrinne f
halfshaft Halbachse f, Achswelle f, Antriebswelle f
halfshaft joint Antriebsgelenk n
halogen headlight Halogenscheinwerfer m
hand Hand f, Hand-
   hand brake Handbremse f
   hand brake cylinder Handbremszylinder m
   hand brake lever Handbremshebel m
   hand brake valve Handbremsventil n
   hand brake warning lamp Handbremsen-Kontrollampe f
   hand control Handverstellung f, Handschaltung f
   hand operation Handbetätigung f

hand pump   Handpumpe f
hand wheel   Handrad n
handle   Griff m, Knopf m, Handgriff m, Schaltgriff m
  handle bar   Lenker m, Lenkstange f *(am Motorrad)*
  door handle   Türgriff m
handling   Handhabung f, Bedienung f, Fahrverhalten n
  handling characteristics   Fahreigenschaften fpl
harness   Leitungssatz m, Kabelsatz m, Kabelbaum m
  harness belt   Hosenträgergurt m
  shoulder harness   Schultergurt m, Sicherheitsgurt m
hatchback   Fahrzeug n mit Heckklappe, Heckklappe f
haul/to   abschleppen
hazard   Unfall m; Panne f
  hazard warning flasher   Pannenwarnblinkgeber m
  hazard warning indicator   Warnblinkerkontrolle f
  hazard warning lamp   Warnblinkleuchte f
  hazard warning signal   Warnblinkleuchte f
head   Kopf m
  head-on crash   Frontalzusammenstoß m
  ball head   Kugelkopf m
  cheese head screw   Zylinderschraube f
  coupling head   Kupplungskopf m
  cylinder head   Zylinderkopf m
  cylinder head gasket   Zylinderkopfdichtung f
  distributor head   Verteilerkappe f
headlamp   Scheinwerferlampe f, Scheinwerfer m
  headlamp casing   Scheinwerfergehäuse n
  headlamp flasher   Lichthupe f
  headlamp mo[u]lding   Scheinwerfereinfassung f
  headlamp reflector   Scheinwerferspiegel m, Reflektor m
  auxiliary headlamp   Zusatzscheinwerfer m
  extra headlamp   Zusatzscheinwerfer m
  main headlamp   Fernscheinwerfer m
  sealed beam headlamp   Sealed-beam-Scheinwerfer m
  (amerikanische Bezeichnung für ein optisches System, bei dem Lichtaustrittscheibe, Reflektor und Glühlampe eine Einheit sind)
  wide-band headlamp   Breitbandscheinwerfer m
headlight   Scheinwerfer m, Fahrzeugscheinwerfer m
  headlight adjustment   Scheinwerfereinstellung f

headlight control   Leuchtweitenregelung f
headlight dimmer switch   Abblendschalter m
headlight flasher   Lichthupe f
headlight insert   Scheinwerfereinsatz m
headlight range adjustment   Leuchtweiteneinstellung f
dipped headlight   Abblendlicht n
halogen headlight   Halogenscheinwerfer m
long-distance beam headlight   Weitstrahler m
headlining   Himmel m des Fahrzeugs
heat   Wärme f, Hitze f
heat absorbing glass   wärmedämmendes Glas n,
Wärmeschutzglas n
heat exchanger   Wärmeübertrager m, Wärmetauscher m
heat insulation   Wärmeisolierung f
heat resistance   Wärmebeständigkeit f
metal heat insulator   Wärmeschutzblech n
heated   beheizt
heated rear window   heizbare Heckscheibe f
heater blower   Heizgebläse n
heater plug   Glühkerze f
heater plug resistor   Glühkerzenwiderstand m
heating   Heizung f
heating element   Heizelement n
heating plug   Glühkerze f
exhaust-operated air heating   Abgasheizung f
heavy-duty truck   Schwerlastwagen m
height adjustment   Höhenjustierung f
helical   schneckenförmig, schraubenförmig
helical gear   Schrägrad n, schrägverzahntes Rad n
helical gearing   Schrägverzahnung f
helical gear shaft   Schraubenradwelle f
helical spring   Schraubenfeder f
hexagon   Sechskant m
hexagon bolt   Sechskantbolzen m
hexagon nut   Sechskantmutter f
hexagon screw   Sechskantschraube f
hexagon socket screw   Innensechskantschraube f, Imbusschraube f
hexagon spanner   Sechskantschlüssel m
high   hoch

high beam Fernlicht n
high-energy ignition Batterie-Hochleistungs-
Kondensator-Zündung f, Thyristorzündung f
high-energy ignition system Batterie-Hochleistungs-
Kondensator-Zündanlage f, Thyristorzündanlage f
high-load enrichment Vollastanreicherung f
high-pressure injection Hochdruckeinspritzung f
high-pressure tire Hochdruckreifen m
high-speed hochtourig
high-speed engine Schnelläufer m, schnellaufender Motor,
hochtouriger Motor m, drehfreudiger Motor m, hoch drehender
Motor m
high-tension ignition Hochspannungszündung f
high-tension ignition cable Hochspannungszündleitung f
hinge Scharnier n
hinge plate Scharnierplatte f
hinged window Ausstellfenster n, Dreieckfenster n
holder Halter m
annular lamp holder Ringfassung f
cable holder Kabelhalter m
lamp holder Lampenfassung f, Lampenhalterung f
holding Halte-
holding bracket Haltebügel m
holding device Haltevorrichtung f
hole circle Lochkreis m
hollow hohl, Hohl-
hollow rivet Hohlniet f
hollow screw Hohlschraube f
hollow shaft Hohlwelle f
homologation allgemeine Betriebserlaubnis f *(Typprüfung)*
honeycomb Wabe f
honeycomb catalyst Wabenkörper-Katalysator m
hood Verdeck n, Motorhaube f
hood bow Verdecksspriegel m
hood catch Haubenverriegelung f, Haubenverschluß m
engine hood Motorhaube f
folding hood Faltdach n, Klappverdeck n
horizontal horizontal, flachliegend
horizontal draught carburetor Flachstromvergaser m

horizontally opposed piston engine Boxermotor m,
  Gegenkolbenmotor m
horn Horn n, Hupe f, Signalhorn n
  horn push button Horndruckknopf m, Hupenknopf m
  multi-tone horn Mehrklanghorn n, Mehrklanghupe f
  signal horn Signalhorn n, Hupe f
hose Schlauch m
  hose clamp (clip) Schlauchbinder m, Schlauchschelle f,
  Schlauchklemme f
  hose coupling Schlauchkupplung f
  hose fitting Schlauchverbindung f
  air vent hose Entlüfterschlauch m
  gasoline hose Benzinschlauch m
  hot air hose Warmluftschlauch m
  insulating hose Isolierschlauch m
hot heiß, warm
  hot air blower Heißluftgebläse n, Warmluftgebläse n
  hot air heating Warmluftheizung f
  hot air hose Warmluftschlauch m
  hot plug Zündkerze f mit niedrigem Wärmewert
  hot wire Hitzdraht m
housing Gehäuse n
  housing cover Gehäusedeckel m
  air filter housing Luftfiltergehäuse n
  clutch housing Kupplungsgehäuse n
  differential housing Differentialgehäuse n
  distributor housing Verteilergehäuse n
  driving housing Antriebsgehäuse n
  flywheel housing Schwungradschüssel f
  plug housing Steckergehäuse n
  thermostat housing Thermostatgehäuse n
hub Nabe f
  hub cap Nabendeckel m, Nabenabdeckung f
  front hub Vorderradnabe f
  wheel hub Radnabe f
  wheel hub cover Radnabendeckel m
hump Hump m *(beim Reifen)*
  hump rim Humpfelge f
hybrid drive Hybridantrieb m

hydraulic hydraulisch
  hydraulic brake system hydraulische Bremsanlage f
  hydraulic coupling Flüssigkeitskupplung f, hydraulische
  Kupplung f
  hydraulic damper hydraulischer Stoßdämpfer m
  hydraulic gear change Drucktölschaltung f
identification plate Typenschild n
idle Leerlauf m
  idle adjustment Leerlaufeinstellung f
  idle air Leerlaufluft f
  idle air bleed Leerlaufluftdüse f, Leerlaufluft-Bohrung f
  idle air control screw Leerlaufluftschraube f,
  Leerlaufluft-Regulierschraube f
  idle air jet Leerlaufluftdüse f, Leerlaufluft-Bohrung f
  idle air screw Leerlaufluftschraube f,
  Leerlaufluft-Regulierschraube f
  idle arm Leerlauf-Steigrohr n, Austrittsarm m für
  Leerlaufkraftstoff m
  idle bleed Leerlaufluftdüse f, Leerlaufluft-Bohrung f
  idle bleed valve Leerlaufluftdüse f
  idle cut-off valve Leerlaufabschaltventil n
  idle jet Leerlaufdüse f
  idle metering jet Leerlaufdüse f, Leerlaufkraftstoffdüse f
  idle mixture Leerlaufgemisch n
  idle mixture adjusting screw Leerlaufgemisch-
  Regulierschraube f, Gemischeinstellschraube f
  idle-mixture control screw Leerlaufgemisch-Regulierschraube f,
  Gemischeinstellschraube f
  idle-mixture setting Leerlaufgemischeinstellung f,
  Gemischeinstellung f
  idle motion stop Leerlaufanschlag m
  idle run Leerlauf m
  idle setting Leerlaufeinstellung f
  idle speed Leerlaufdrehzahl f
  idle speed adjusting screw Leerlaufeinstellschraube f,
  Leerlaufbegrenzungsschraube f
  idle speed adjustment Leerlaufeinstellung f
  idle stabilization Leerlaufstabilisierung f
  idle system Leerlaufsystem n

idle tube Leerlauf-Steigrohr n, Austrittsarm m für
Leerlaufkraftstoff
idler Zwischenrad n, Spannrad n, Spannrolle f, Vorgelegerad n
  idler arm Lenkzwischenhebel m
  idler gear Zwischenrad n, Vorgelegerad n
  idler pulley Leerlaufscheibe f, Spannrolle f, Umlenkrolle f
  idler shaft Vorgelegeachse f
  reverse idler gear Rücklaufrad n
idling Leerlauf m
  idling arm Leerlauf-Steigrohr n, Austrittsarm m für
  Leerlaufkraftstoff
  idling characteristic Leerlaufkennlinie f
  idling feed orifice Leerlaufbohrung f
  idling jet Leerlaufdüse f
  idling metering jet Leerlaufdüse f, Leerlaufkraftstoffdüse f
  idling mixture Leerlaufgemisch n
  idling speed Leerlaufdrehzahl f
  idling system Leerlaufsystem n
  idling tube Leerlauf-Steigrohr n, Austrittsarm m für
  Leerlaufkraftstoff
ignitable zündfähig
ignite/to zünden
ignition Zündung f
  ignition accelerator Zündbeschleuniger m
  ignition advance Zündzeitpunktvorgabe f
  ignition aid Zündhilfsmittel n, Starthilfe f
  ignition cable Zündleitung f, Zündkabel n
  ignition cable plug Zündleitungsstecker m
  ignition cam Unterbrechernocken m
  ignition capacitor Zündkondensator m
  ignition circuit Zündschaltung f, Zündleitung f
  ignition coil Zündspule f
  ignition contact breaker Zündunterbrecher m
  ignition control Zündzeitverstellung f
  ignition cycle Verbrennungshub m
  ignition device Zündeinrichtung f
  ignition distributor Zündverteiler m
  ignition failure Zündaussetzer m
  ignition impulse collector Zündimpulsabnehmer m

ignition key   Zündschlüssel m
ignition lock   Zündschloß n
ignition order   Zündfolge f
ignition point   Zündzeitpunkt m
ignition setting   Zündeinstellung f
ignition starter switch   Zündanlaßschalter m
ignition stroke   Verbrennungshub m, Verbrennungstakt m
ignition switch   Magnetschalter m, Zündschalter m
ignition system   Zündanlage f
ignition timer   Zündverteiler m
ignition timing   Zündverstellung f, Zündzeitpunktverstellung f
ignition timing line   Zündverstellinie f
ignition voltage   Zündspannung f
ignition warning light   Zündkontrollampe f
advance[d] ignition   Vorzündung f
applied ignition   Fremdzündung f
battery-coil ignition   Batteriezündung f
battery ignition   Batteriezündung f
early ignition   Frühzündung f, Vorzündung f
glow ignition   Glühzündung f
high-tension ignition cable   Hochspannungszündleitung f
low-tension ignition circuit   Niederspannungszündleitung f
premature ignition   Frühzündung f
retarded ignition   Spätzündung f
spark ignition   Fremdzündung f, Funkenzündung f
transistor ignition   Transistorzündung f
vacuum ignition adjustment   Unterdruckzündverstellung f
variable ignition   Zündpunktverstellung f
illumination   Beleuchtung f
imbalance   Unwucht f
imitation leather   Kunstleder n
impact   Aufprall m, Stoß m
  impact energy   Aufprallenergie f
  impact sensor   Aufprallsensor m
impeller   Laufrad n, Schaufelrad n, Ventilatorrad n
independent   unabhängig, Einzel-
  independent suspension   Einzelradaufhängung f
  independent wheel suspension   Einzelradaufhängung f
indicator   Anzeiger m, Anzeigegerät n, Anzeigeinstrument n

indicator control switch   Winkerschalter m
indicator lamp   Anzeigelampe f, Kontrollampe f
indicator light   Anzeigeleuchte f, Kontrollampe f
indicator switch   Fahrtrichtungsanzeigeschalter m, Winkerschalter m
direction indicator   Fahrtrichtungsanzeiger m, Winker m
flasher indicator switch   Blinkschalter m
fuel level indicator   Kraftstoffanzeiger m
main beam indicator   Fernlichtkontrollampe f
mileage indicator   Meilenzähler m
indirect injection   indirekte Einspritzung f, Vorkammereinspritzung f *(beim Dieselmotor)*
indoor   Innen-
indoor lamp   Innenleuchte f
indoor lighting   Innenbeleuchtung f
induction   1. Induktion f *(Elektrik)*; 2. Ansaug-
induction air   Ansaugluft f
induction air filter   Ansaugluftfilter n,m
induction coil   Induktionsspule f
induction manifold   Einlaßkrümmer m, Ansaugkrümmer m, Saugrohr n
induction pipe   Ansaugrohr n, Einlaßrohr n
induction port   Ansaugkanal m, Einlaßkanal m, Saugkanal m
induction stroke   Ansaughub m, Einlaßhub m
inertia   Massenträgheit f, Trägheit f
inertia brake   Auflaufbremse f
inertia retractor   Verzögerungsaufrolleinrichtung f *(Sitzgurt)*
infinitely   stufenlos
infinitely variable   stufenlos regelbar
infinitely variable speed transmission   stufenloses Getriebe n
inflation   Füll-
inflation pressure   Fülldruck m *(Reifen)*
tire-inflation pressure   Reifeninnendruck m, Fülldruck m
infrared headlamp   Infrarotscheinwerfer m
inhibitor   Inhibitor m, Additiv n
initial   Anfangs-, Erst-
initial fill-up   Erstfüllung f *(Betriebsmittel vom Hersteller)*
initial pressure   Anfangsdruck m *(Lader)*
initial speed   Anfangsdrehzahl f, Anfangsgeschwindigkeit f

injection Einspritzung f
  injection advance Einspritzvoreilung f, Voreinspritzung f
  injection advance device Einspritzversteller m
  injection engine Einspritzmotor m
  injection lag Einspritzverzögerung f
  injection line Einspritzleitung f
  injection nozzle Einspritzdüse f
  injection period Einspritzdauer f, Öffnungsdauer f,
  Ventilerhebungszeit f
  injection pipe Einspritzleitung f
  injection port Einspritzöffnung f
  injection pressure Einspritzdruck m, Öffnungsdruck m
  injection system Einspritzanlage f
  injection timer Spritzversteller m
  injection timing Einspritzverstellung f *(Dieselmotor)*,
  Spritzzeitpunktverstellung f
  injection valve Einspritzventil n
  direct injection Zylindereinspritzung f
  fuel injection pump Einspritzpumpe f
injector Einspritzdüse f
  injector nozzle Einspritzdüse f
  injector tube Einspritzrohr n, Spritzrohr n
inlet Eingang m, Einlaß m
  inlet air Ansaugluft f
  inlet cam Einlaßnocken m
  inlet camshaft Einlaßnockenwelle f, Nockenwelle f für Einlaßventile
  inlet connection Einlaufstutzen m *(Kühler oder Wasserpumpe)*
  inlet duct Einlaßkanal n, Ansaugkanal m
  inlet horn Einlaßtrichter m, Ansaugtrichter m
  inlet line Einlaßleitung f
  inlet manifold Einlaßkrümmer m, Ansaugkrümmer m, Saugrohr n
  inlet manifold pressure Druck m im Einlaßkrümmer, Druck m im
  Ansaugkrümmer, Druck m im Saugrohr, Saugrohrdruck m
  inlet pipe Ansaugrohr n, Saugrohr n, vorderes Auspuffrohr n;
  Einlaufstutzen m
  inlet port Einlaßkanal m, Ansaugkanal m, Einlaß-Steueröffnung f
  inlet trumpet Einlaßtrichter m, Ansaugtrichter m
  inlet tube Ansaugrohr n, Saugrohr n
  inlet valve Einlaßventil n, Ansaugventil n

inlet valve opening   Einlaßbeginn m, Einlaßöffnung f
air inlet   Lufteinlaß m
air inlet tube   Lufteintrittstutzen m, Einlaßöffnung f
in-line   Reihen-
  in-line engine   Reihenmotor m
  in-line pump   Reiheneinspritzpumpe f
  in-line type injection pump   Reiheneinspritzpumpe f
inner   Innen-
  inner panel   Innenblech n
  inner skin panel   Innenblech n
  inner tube   Luftschlauch m *(Reifen)*
  inner-tube valve   Schlauchventil n
input   Eingang m
  input cylinder   Geberzylinder m
  input shaft   Eingangswelle f, Antriebswelle f
  input torque   Eingangsdrehmoment n
  current input   Stromaufnahme f
  power input   Leistungsaufnahme f
insert   Einlage f, Einsatz m
  code insert   Codiereinsatz m
  filter insert   Filtereinsatz m
  flat insert tongue   Flachsteckzunge f
  headlight insert   Scheinwerfereinsatz m
  joint insert   Dichtungseinsatz m
  metal insert   Blecheinlage f
  spark plug insert   Kerzeneinsatz m
  spring steel insert   Federstahleinlage f
  threaded insert   Gewindeeinsatz m
  valve insert   Ventileinsatz m
inside   Innen-
  inside door panel   Türinnenblech n
  inside fittings   Innenausstattung f
  inside lining   Innenauskleidung f, Innenverkleidung f
  inside mirror   Innenspiegel m, Innenrückspiegel m
  inside panel   Innenblech n
  inside rear-view mirror   Innenrückspiegel m, Innenspiegel m
  inside roof lining   Himmel m *(des Fahrzeugs)*
  inside shoe brake   Innenbackenbremse f
inspection   Prüfung f

installation   Aggregat n, Anlage f, Einbau m
  installation dimension   Einbaumaß n
instruction   Anweisung f, Anleitung f
  operating instructions   Bedienungsanleitung f
instrument   Instrument n, Anzeigegerät n
  instrument board   Armaturenbrett n
  instrument cluster   Kombiinstrument n, Instrumentengruppe f
  instrument lamp   Armaturenleuchte f
  instrument panel   Instrumententafel f, Armaturenbrett n
  measuring instrument   Meßinstrument n
  testing instrument   Prüfgerät n
insulate/to   isolieren
insulating   Isolier-
  insulating cap   Isolierkappe f; Polkappe f *(Batterie)*
  insulating hose   Isolierschlauch m
  insulating material   Isoliermaterial n, Isolierstoff m, Dämmaterial n
  insulating sleeve   Isolierhülse f
  insulating tape   Isolierband n
  insulating tube   Isolierrohr n
insulation   Isolierung f
  flexible insulation tubing   Isolierschlauch m
  sound insulation mat   Schalldämm-Matte f
  strip the insulation/to   abisolieren
insulator   Isolator m, Isolierkörper m
  metal heat insulator   Wärmeschutzblech n
intake   Eintritt, Ansaug-
  intake air   Ansaugluft f
  intake air heater   Ansaugluftvorwärmer m, Vergaserluftvorwärmer m
  intake cam   Einlaßnocken m
  intake camshaft   Einlaßnockenwelle f
  intake hose   Ansaugschlauch m, Luftsaugschlauch m
  intake line   Einlaßleitung f, Ansaugleitung f
  intake manifold   Einlaßkrümmer m, Ansaugkrümmer m, Saugrohr n
  intake manifold pressure   Druck m im Einlaßkrümmer, Druck m im Ansaugkrümmer, Druck m im Saugrohr, Saugrohrdruck m
  intake muffler   Ansaugschalldämpfer m

intake pipe   Ansaugrohr n, Saugrohr n
intake port   Einlaßkanal m, Ansaugkanal m
intake side   Einlaßseite f
intake silencer   Ansaugschalldämpfer m
intake stroke   Saughub m, Ansaughub m, Ansaugtakt m,
Einlaßhub m
intake trumpet   Einlaßtrichter m, Ansaugtrichter m
intake valve   Einlaßventil n, Ansaugventil n
air intake tube   Luftansaugrohr n
integral   zusammenhängend *(konstruktiv)*, selbsttragend
integral body   selbsttragende Karosserie f
integral frame construction   rahmenlose Bauweise f, voll
selbsttragender Aufbau m
intercooler   Ladeluftkühler m
intercooling   Ladeluftkühlung f
interference   Störung f, Überlagerung f, Beeinflussung f
interference suppression   Entstörung f
interference suppression equipment   Entstörbausatz m
anti-interference cable   Entstörkabel n
anti-interference capacitor   Entstörkondensator m
intermediate   Zwischen-
intermediate bearing   Zwischenlager n
intermediate gear   Zwischenrad f, Vorgelege n
intermediate housing   Zwischengehäuse n
intermediate layer   Zwischenlage f
intermediate shaft   Vorgelegewelle f, Zwischenwelle f
intermediate transmission   Zwischengetriebe n
intermittent   intermittierend, aussetzend
intermittent wipers   intermittierende Scheibenwischer mpl
internal   Innen-
internal combustion engine   Verbrennungsmotor m
internal diameter   Innendurchmesser m
internal gearing   Innenverzahnung f
internally toothed belt   innenverzahnter Zahnriemen m
internally ventilated disk brake   innenbelüftete Scheibenbremse f
interrupt/to   unterbrechen, ausschalten
interrupter   Unterbrecher m *(Zündung)*; Schalter m
door contact interrupter   Türkontaktschalter m
interruption   Unterbrechung f

intersection   Kreuzung f, Straßenkreuzung f
inverted engine   Motor m mit hängenden Ventilen, hängender
  Motor m
jack/to   heben *(Wagen mit Heber)*
jack   Wagenheber m
  lifting jack   Wagenheber m
jacket   Umhüllung f, Mantel m
  cylinder jacket   Zylindermantel m
  exhaust-jacket carburetor   Vergaser m mit Auspuffheizmantel
jacking point   Ansatzstelle f für Wagenheber
jam/to   klemmen, festsitzen *(Kolben)*
jaw   Klemmbacke f, Backe f
  jaw-clutch   Klauenkupplung f
  clamping jaw   Klemmbacke f, Spannbacke f
jet   Düse f
  jet carburetor   Düsenvergaser m, Zerstäubervergaser m
  air correction jet   Ausgleichluftdüse f, Luftkorrekturdüse f
  ball jet   Kugeldüse f
  change-over jet   Übergangsdüse f
  double jet   Doppeldüse f
  easy start jet   Startdüse f
  idle (idling) jet   Leerlaufdüse f
  overflow jet   Überlaufdüse
  pilot jet   Voreinspritzdüse f, Vordüse f
  single jet   Einfachdüse f
joint   Gelenk n, Verbindung f, Fuge f
  joint ball   Gelenkkugel f
  joint coupling   Gelenkkupplung f
  joint face   Dichtfläche f *(Motorgehäuse)*
  joint insert   Dichtungseinsatz m
  joint ledge   Abdichtleiste f
  joint lever   Gelenkhebel m
  joint piece   Gelenkstück n
  joint pin   Gelenkzapfen m
  joint profile   Dichtprofil n
  joint washer   Dichtscheibe f, Flachdichtring m
  axial joint   Axialdichtung f
  ball joint   Kugelgelenk n
  ball joint socket   Kugelgelenkgehäuse n

cardan joint   Kardangelenk n
circumferential joint   Ringdichtung f
claw joint   Klauengelenk n
pipe joint   Rohrdichtung f, Rohrverschraubung f
radial joint   Radialdichtung f
rubber joint   Gummidichtung f
slip universal joint   Gleitkreuzgelenk n
universal joint   Gelenkkupplung f, Kreuzgelenk n, Kardangelenk n
universal-joint yoke   Kreuzgelenkgabel f
journal   Lagerzapfen m, Zapfen m
  axle journal   Lagerhals m
  neck journal bearing   Halslager n
jump spark   Überspringfunke m
jumper   Überbrückungs-
  jumper cable (lead)   Starthilfekabel n
junction   Abzweigung f
  junction box   Anschlußdose f, Abzweigdose f
kerb   Bordstein m, Randstein m *(US: curb)*
  kerb feeler   Bordsteinfühler m, Reifenwächter m
key   Keil m, Schlüssel m, Taste f
  key groove (slot)   Keilnut f
  contact key   Kontakttaster m
  ignition key   Zündschlüssel m
  safety key   Sicherungskeil m
  tightening key   Spannschlüssel m
keyboard   Tastatur f
kick   Stoß m, Schlag m, Schlagen n
  kick-down   Kickdown m, Übergas n *(automatisches Getriebe)*
  kick starter   Kickstarter m *(beim Motorrad)*
kingpin   Spurzapfen m, Achsschenkelbolzen m
  kingpin angle   Radspreizungswinkel m
  kingpin offset   Lenkrollradius m
  kingpin steering assembly   Achsschenkellenkeinrichtung f
kit   Satz m, Teilesatz m
  tool kit   Werkzeugkasten m, Werkzeugsatz m
knob   Knopf m
  control knob   Betätigungsknopf m
  starting knob   Starterknopf m
knock/to   klopfen, nageln

knock Schlag m, Klopf-
  knock inhibitor Klopfbremse f
  knock limit Klopfgrenze f
  knock-proof klopffest
  knock rating Oktanzahl f, Klopffestigkeit f
  knock resistance Klopffestigkeit f
  knock suppressor Klopfbremse f
  knock tendency Klopfneigung f
  combustion knock Klopfen n, Verbrennungsklopfen n
knocking Klopfen n, Kolbenklopfen n
  knocking limit Klopfgrenze f
  knocking noise Klopfgeräusch n
knockless klopffrei, klopffest
knuckle Gelenk n
  knuckle pin Achsschenkelbolzen m
  steering knuckle Lenkachsenschenkel m, Achsschenkel m
  steering knuckle pivot Vorderachszapfen m
knurled gerändelt
  knurled grip Rändelgriff m
  knurled head screw Rändelschraube f
  knurled nut Rändelmutter f
lacquer Lack m
lag Verzug m, Verzögerung f
lamella Lamelle f
  lamella clutch Lamellenkupplung f
laminated beschichtet; mehrlagig
  laminated glass Verbundglas n
  laminated safety glass Verbundsicherheitsglas n, Verbundglas n
  laminated spring Blattfeder f
lamp Lampe f; Leuchte f
  lamp holder Lampenfassung f, Lampenhalterung f
  lamp socket Glühlampenfassung f, Lampensockel m,
  Lampenfassung f
  adjustable spot lamp Suchscheinwerfer m
  back-up lamp Rückfahrleuchte f, Rückfahrscheinwerfer m
  ceiling lamp Deckenleuchte f, Deckenlampe f
  combined flash-tail lamp kombinierte Blink-Schlußleuchte f
  dash lamp Armaturenleuchte f
  dashboard lamp Armaturenleuchte f

direction-indicator control lamp Blinkerkontrolleuchte f
driving lamp Fernscheinwerfer m, Weitstrahler m
flasher back-up lamp Blink-Rückfahrleuchte f
flasher lamp Blinkleuchte f, Blinker m
fog lamp Nebellampe f, Nebelscheinwerfer m
hand brake warning lamp Handbremsen-Kontrollampe f
indicator lamp Anzeigelampe f, Kontrollampe f
indoor lamp Innenleuchte f
instrument lamp Armaturenleuchte f
licence plate lamp Kennzeichenleuchte f, Nummernschildleuchte f
main beam warning lamp Fernlichtkontrollampe f
panel lamp Armaturenleuchte f
parking lamp Parkleuchte f
position lamp Positionslampe f
reading lamp Leseleuchte f
rear flasher lamp Heckblinkleuchte f, Schlußblinkleuchte f
rear fog-lamp Nebelschlußleuchte f
rear lamp Schlußleuchte f, Heckleuchte f
reversing lamp Rückfahrscheinwerfer m
safety lamp Warnleuchte f
side-flasher lamp Begrenzungsblinkleuchte f
side lamp Begrenzungsleuchte f, Seitenleuchte f, Positionsleuchte f
side-marker lamp Seiten*markierungs*leuchte f,
Begrenzungsleuchte f
spot lamp Suchscheinwerfer m, Weitstrahler m
spotlight lamp Suchscheinwerfer m
stop lamp Bremsleuchte f
tail lamp Schlußleuchte f
tell-tale lamp Ladekontrollampe f
warning lamp Kontrollampe f
lane Fahrbahn f; Fahrspur f
  lane changing Fahrspurwechsel m
  passing lane Überholspur f
lap and shoulder belt Dreipunktgurt m
lap belt Beckengurt m
latch Klinke f, Riegel m, Verriegelung f
  latch striker Schließplatte f *(Tür)*
  hood latch Verdeckriegel m
  locking latch Sperrzunge f

lateral   seitlich, Quer-
  lateral acceleration   Querbeschleunigung f, Seitenbeschleunigung f
  lateral force   Seitenführungskraft f, Spurhaltungskraft f
  lateral tire run-out   Seitenschlag m, Planlaufabweichung f *(Reifen)*
  lateral wheel run-out   Seitenschlag m, Planlaufabweichung f *(Rad)*
layer   Lage f *(Reifen)*
  barrier layer   Sperrschicht f *(Reifen)*
  intermediate layer   Zwischenlage f *(Reifen)*
lead   Leitung f; Blei n
  battery lead   Batterieleitung f
  earth lead   Masseleitung f
  short-circuit lead   Kurzschlußleitung f
leaded fuel   verbleiter Kraftstoff m
lead-free fuel   unverbleiter Kraftstoff m, bleifreier Kraftstoff m
leaf   Federblatt n
  leaf spring   Blattfeder f
  spring leaf   Federblatt n
leak air   Falschluft f
leakage   Leck n, Undichtheit f
  leakage current   Kriechstrom m
  leakage fuel   Leckkraftstoff m, Leckage f
  leakage rate   Leckrate f, Leckmenge f
  leakage test   Dichtheitsprüfung f
lean   mager *(Gemisch)*
  lean mixture   mageres Gemisch n
leather   Leder n, Lederauflage n
  leather bellows   Lederbalg m
  leather sleeve   Ledermanschette f, Lederhülse f
  imitation leather   Kunstleder n
ledge   Leiste f
  joint ledge   Abdichtleiste f
  profile ledge   Profilleiste f
left-hand   links, links-, linksgängig
  left-hand drive   Linkslenkung f
  left-hand rotation   Linkslauf m, Linksdrehung f
  left-hand steering   Linkslenkung f
  left-hand turn   Linksabbiegen n
length   Länge f
  length of brake path   Bremsweg m

  spark length   Funkenlänge f
lens   Linse f, Lichtaustrittscheibe f, Streuscheibe f,
level   Ebene f, Höhe f, Niveau n, Stand m, Spiegel m
  level adjustment   Höheneinstellung f, Niveauregulierung f
  (Federung)
  level control   Höheneinstellung f, Niveauregulierung f *(Federung)*
  level ga[u]ge   Niveaumeter n
  oil level tube   Ölstandsrohr n
lever   Hebel m
  lever arm   Hebelarm m
  adjusting lever   Einstellhebel m
  control lever   Schalthebel m, Betätigungshebel m
  disengaging lever   Ausrückhebel m
  foot lever   Fußhebel m
  gear lever   Schalthebel m, Getriebeschalthebel m
  gear-shift lever   Schalthebel m, Handschalthebel m
  gear-shift lever shaft   Schaltwelle f
  hand brake lever   Handbremshebel m
  operating lever   Betätigungshebel m
  preselector lever   Vorwählhebel m
  release lever   Auslösehebel m
  reversing lever   Umlenkhebel m
  rocker lever   Kipphebel m, Schlepphebel m
  speed-change lever   Gangwechselhebel m, Schalthebel m
  steering box lever   Lenkstockhebel m
  timing lever   Verstellhebel m
licence   Zulassung f, Zulassungsschein m; Fahrerlaubnis f
  licence lamp   Nummernschildleuchte f, Kennzeichenleuchte f
  licence plate   Nummernschild n, Kennzeichen n
  licence plate lamp   Kennzeichenleuchte f, Nummernschildleuchte f
  driving licence   Fahrerlaubnis f, Führerschein m
lid   Deckel m
lifter   Ausheber m, Nocken m
  window lifter   Fensterheber m
lifting platform   Hebebühne f
light   1. Licht n, Leuchte f, Lampe f; 2. leicht
  light alloy   Leichtmetallegierung f
  light alloy wheel   Leichtmetallrad n
  light exit   Lichtaustritt m

light metal   Leichtmetall n
light metal wheel   Leichtmetallrad n
light switch   Lichtschalter m
light transmission   Lichtdurchlässigkeit f
light unit   Scheinwerfereinsatz m
light-weight construction   Leichtbauweise f
back-up light   Rückfahrlicht n
dim (dimmed, dimming) light   Abblendlicht n
driving light   Fernlicht n
dynamo charging light   Ladekontrollicht n
exit area of light   Lichtaustrittfläche f
exit of light   Lichtaustritt m
flasher warning light   Blinkkontrollampe f
flashing light   Blinklicht n
indicator light   Anzeigeleuchte f, Kontrollampe f
intensity of light   Helligkeit f
long-distance light   Fernlicht n
main light   Fernlicht n, Hauptlicht n
map light   Leseleuchte f
marker light   Begrenzungslampe f
parking light   Parklicht n, Standlicht n
passing light   Abblendlicht n
petrol warning light   Benzinkontrolleuchte f
rear light   Rücklicht n, Schlußlicht n
side light   Begrenzungslicht n, Seitenleuchte f
stop light   Bremslicht n, Stopplicht n
tail light   Schlußlicht n, Rücklicht n
warning light   Kontrolleuchte f
widespread light   Breitstrahler m
lighter   Anzünder m
  cigar lighter   Zigarrenanzünder m
lighting   Beleuchtung f
  indoor lighting   Innenbeleuchtung f
  number-plate lighting   Kennzeichenbeleuchtung f
  panel lighting   Armaturenbrettbeleuchtung f
limit   Höchstgrenze f, Grenzwert m
  limit switch   Endschalter m, Endausschalter m
  speed limit   Geschwindigkeitsbeschränkung f, zugelassene
  Höchstgeschwindigkeit f

line Leitung f, Linie f
 air relief line Entlüfterleitung f
 connecting line Verbindungsleitung f
 feed line Zuleitung f
 flexible tube line Schlauchleitung f
 fuel-supply line Kraftstoffleitung f
 leakage oil line Lecköleitung f
 pipe line Rohrleitung f
 short circuit line Kurzschlußleitung f
liner Laufbüchse f, Büchse f, Zylinderbüchse f; Futter n,
 Auskleidung f
 cylinder liner Zylinderbüchse f
lining Belag m, Überzug m, Verkleidung f, Innenauskleidung f
 brake lining Bremsbelag m
 clutch lining Kupplungsbelag m
 friction lining Reibbelag m
 hand brake lining Bandbremsbelag m
 head lining Himmel m des Fahrzeugs
 inside roof lining Himmel m des Fahrzeugs
link Gelenk n; Lenker m; Achsstrebe f
 link rod Gelenkstange f
 link strut Lenkerstange f
 double transverse link Doppelquerlenker m
 drag link Lenkspurhebel m, Lenkstange f
 trailing link Längslenker m
 transverse link Querlenker m
linkage Gestänge n
 linkage brake Gestängebremse f
 windscreen wiper linkage Scheibenwischergestänge n
 wiper linkage Wischergestänge n
liquefied petrolium gas Flüssiggas n
live axle Starrachse f, Antriebsachse f
load Belastung f, Beladung f, Kraft f
 load capacity Belastungsfähigkeit f, Belastbarkeit f
 load characteristic Belastungskennlinie f
 load compensation Belastungsausgleich m
 load cycle Lastspiel n, Lastwechsel m
 load-dependent beladungsabhängig, lastabhängig
 load-dependent brake force proportioning

    device lastabhängiger Bremskraftverteiler m
    load floor Radfläche f
    load sensitive lastempfindlich, belastungsempfindlich
    full load Vollast f
    partial load Teillast f
loading Lade-
    loading area Ladefläche f
    low-loading trailer Tiefbettanhänger m
lock Schloß n, Sperre f, Verriegelung f, Sicherung f
    lock angle Einschlagwinkel m *(Räder)*
    lock bolt Sperriegel m
    lock cylinder Schloßzylinder m
    lock nut Gegenmutter f, Kontermutter f, Verschlußmutter f
    lock pawl Schloßklinke f
    lock ring Sicherungsring m
    lock screw Verschlußschraube f
    lock striker Schließkeil m
    lock washer Sicherungsscheibe f
    angle of lock Ausschlagwinkel m *(Lenkung)*, Einschlagwinkel m *(Räder)*
    bajonet lock Bajonettverschluß m
    differential lock Ausgleichsperre f, Differentialsperre f
    door lock Türschloß n
    door lock cylinder Türschließzylinder m
    ignition lock Zündschloß n
    radiator lock Kühlerverschluß m
    safety lock Sicherheitsschloß n
    toothed lock washer Zahnscheibe f
    vapo[u]r lock Blasenbildung f *(in Kraftstoffleitungen)*
locking Verschluß-, Sperr-, Schließ -
    locking cap Verschlußkappe f
    locking latch Sperrzunge f
    locking pawl Schließraste f, Sperrklinke f
    locking pin Arretierstift m, Haltestift m
    locking plug Verschlußstopfen m
    locking ring Sicherungsring m, Sperring m
    locking screw Verschlußschraube f
lock-up indicator Blockieranzeiger m
long lang, fern, weit

long-distance beam   Weitstrahler m
long-distance haulage/hauling   Güterfernverkehr m, Fernverkehr m
long-distance light   Fernlicht n
long-distance traffic   Fernverkehr m
longitudinal   Längs-
  longitudinal control arm   Längslenker m
  longitudinal link   Längslenker m
  longitudinal member   Längsträger m
  longitudinal shaft   Längswelle f
loom   Kabelstrang m, Kabelbaum m
loop   Schlaufe f
  loop joint   Rundschnurdichtung f
  loop-ring   Rundschnurring m
  loop scavenging   Umkehrspülung f
loose connection   Wackelkontakt m
lorry   Lastkraftwagen m
loudspeaker   Lautsprecher m
  loudspeaker cable   Lautsprecherkabel n
low   niedrig, tief; Tief-, Nieder-
  low beam   Abblendlicht n
  low-built chassis   Niederrahmenfahrgestell n
  low-emission engine   Motor m mit geringem Schadstoffausstoß, schadstoffarmer Motor m
  low-fuel indicator   Kraftstoffreserveanzeiger f
  low-gear   niedriger Gang m, Anfahrgang m
  low-loader   Tiefladeanhänger m, Tieflader m
  low-loading trailer   Tiefbettanhänger m
  low-pressure box   Unterdruckdose f
  low-pressure line   Unterdruckleitung f
  low-pressure tire   Niederdruckreifen m
  low-profile tire   Niederquerschnittsreifen m
  low-section tire   Niederquerschnittsreifen m
  low-voltage ignition circuit   Niederspannungszündleitung f
lower beam   Abblendlicht n
lubricant   Schmiermittel n
lubricated petrol   Kraftstoff-Öl-Gemisch n, Zweitaktgemisch n
lubricating   Schmier-
  lubricating film   Schmierfilm m

lubricating grease Schmierfett n
lubricating gun Schmierpresse f, Fettpresse f
lubricating oil Schmieröl n
lubricating point Schmierstelle f
lubrication Schmierung f
lubrication nipple Schmiernippel m
lubrication piping Schmierleitung f
central chassis lubrication Zentralschmierung f
continuous lubrication Dauerschmierung f
petrol lubrication Gemischschmierung f, Zweitaktschmierung f
ring lubrication Ringschmierung f
lug Öse f, Nase f; Stollen m *(Reifen)*
lug cam Anschlagnocken m
gear-shift lug Schaltbock m
terminal lug Anschlußfahne f
luggage Gepäck n, Koffer m
luggage boot Kofferraum m
luggage capacity Kofferraumvolumen n
luggage compartment Kofferraum m
luggage compartment lid Kofferraumdeckel m
luminous leuchtend, Licht-
luminous efficiency Lichtausbeute f
luminous intensity Lichtstärke f
luxury limousine Luxuslimousine f
magnetic magnetisch
magnetic brake Magnetbremse f
magnetic switch Magnetschalter m
main Haupt-
main air reservoir Hauptluftbehälter m *(Bremse)*
main beam Fernlicht n, aufgeblendetes Licht n
main beam indicator Fernlichtkontrollampe f
main beam warning lamp Fernlichtkontrollampe f
main bearing Hauptlager n
main brake cylinder Hauptbremszylinder m
main chamber Hauptkammer f, Hauptbrennraum m
main circuit Hauptstromkreis m
main current switch Hauptstromschalter m
main jet Hauptdüse f, Vollastdüse f
main journal Hauptlagerzapfen m

main light   Fernlicht n, Hauptlicht n
main shaft   Getriebehauptwelle f
  battery main switch   Batteriehauptschalter m
maintenance   Wartung f, Instandhaltung f
  maintenance-free   wartungsfrei
make-and-break contact   Unterbrecherkontakt m
male   Steck-, Stift-
  male plug   Stecker m
  male socket   Steckerbuchse f
mandrel   Dorn m
  mounting mandrel   Montagedorn m
manifold   Krümmer m, Verteiler m, Sammelleitung f
  manifold injection   Saugrohreinspritzung f
  exhaust manifold   Auspuffkrümmer m
  induction manifold   Ansaug*sammel*leitung f, Einlaßkrümmer m
manometer   Manometer n, Druckmeßgerät n
manual   1. manuell, Hand-; 2. Bedienungsanleitung f, Handbuch n
  manual gearbox   Schaltgetriebe n, Handschaltgetriebe n
  manual selection   Handschaltung f
  manual shift   handbetätigter Schaltvorgang m, manueller Gangwechsel m
  manual steering gear   mechanisches Lenkgetriebe n
  manual throttle control   Handgasbetätigung f
  manual transmission   Schaltgetriebe n, Handschaltgetriebe n
map light   Leseleuchte f
mark   Marke f, Markierung f
  adjusting mark   Einstellmarke f
  flywheel timing mark   Schwungradmarke f
  setting mark   Einstellmarke f
marker   Markierungs-
  marker light   Begrenzungslampe f
  side-marker lamp   Seiten[markierungs]leuchte f, Begrenzungsleuchte f
masked headlamp   Tarnscheinwerfer m
mass   Masse f
  centrifugal mass   Schwungmasse f
master   Haupt-
  master cylinder   Hauptbremszylinder m, Geberzylinder m
  master gear   großes Differentialantriebskegelrad n

mat Matte f
  floor mat Bodenmatte f, Fußbodenmatte f
  sound insulation mat Schalldämm-Matte f
material testing Werkstoffprüfung f, Materialprüfung f
maximum Maximal-, Höchst-
  maximum load Höchstbelastung f
  maximum load rating zulässige Belastung f
  maximum output Höchstleistung f, maximale Leistung f
  maximum permissible speed zulässige Höchstgeschwindigkeit f
  maximum permissible weight zulässige Gesamtmasse f, maximal
  zulässige Masse f
  maximum power Höchstleistung f, maximale Leistung f
  maximum speed Höchstdrehzahl f, Höchstgeschwindigkeit f
  maximum torque Höchstdrehmoment n, maximales
  Drehmoment n
mean effective pressure Mitteldruck m, mittlerer Arbeitsdruck m
measurement Messung f
measuring Meß -
  measuring instrument Meßinstrument n
  measuring range Meßbereich m
mechanical mechanisch
  mechanical fuel injection mechanische Kraftstoffeinspritzung f
  mechanical injection mechanische Einspritzung f
  mechanical mixture control mechanische
  Kraftstoffeinspritzung f, mechanische Gemischregelung f
  mechanical steering mechanische Lenkung f
  mechanical steering gear mechanisches Lenkgetriebe n
  mechanical stress mechanische Belastung f
mechanism Mechanik f, Vorrichtung f, Getriebe n
  gear selector mechanism Gangwähler m
  gear-shift mechanism Gangschaltung f
  power-steering mechanism Servolenkung f, Servolenkgetriebe n
  window mechanism Fensterheber m
member Teil n, Träger m
  longitudinal member Längsträger m
  side member Längsträger m
  tubular cross member Rohrquerträger m
metal Metall n
  sheet metal Blech n

sintered metal   Sintermetall n
metallic paint finish   Metalliclackierung f
meter   Meßinstrument n, Zähler m
  torque meter   Drehmomentschlüssel m
metering   Dosier-
  metering needle   Düsennadel f *(zur Kraftstoffdosierung)*
  metering plunger   Steuerkolben m
  metering port (slit)   Steuerschlitz m
  metering valve   Gemischregler m
metric thread   metrisches Gewinde n
microswitch   Mikroschalter m
mid-engined car   Fahrzeug n mit Mittelmotor
mileage   gefahrene Wegstrecke f, zurückgelegter Weg m *(auf Meilen bezogen)*
  mileage indicator   Anzeigegerät n für den insgesamt zurückgelegten Weg *(auf Meilen bezogen)*
  mileage recorder   Fahrtenschreiber m
  fuel mileage   Kraftstoffergiebigkeit f *(auf Meilen bezogen)*
mirror   Spiegel m
  mirror bracket   Spiegelhalter m, Spiegelhalterung f
  dimming mirror   Abblendspiegel m
  rear mirror   Rück*blick*spiegel m, Außenspiegel m
  rear view mirror   Rück*blick*spiegel m
misfire   Fehlzündung f, Zündaussetzer m, Aussetzer m
miss/to   aussetzen *(Motor)*
mixing   Misch-
  mixing chamber   Mischkammer f *(Vergaser)*
  mixing ratio   Mischungsverhältnis n
  mixing tube   Mischrohr n *(Vergaser)*
mixture   Mischung f, Gemisch n
  mixture admission   Gemischzuführung f
  mixture control   Gemischregelung f
  mixture control screw   Gemischeinstellschraube f, Leerlaufgemisch-Regulierschraube f
  mixture control unit   Gemischregler m
  mixture distribution   Gemischverteilung f
  mixture enrichment   Gemischanreicherung f, Kraftstoffanreicherung f
  mixture formation   Gemischbildung f

mixture preheating Gemischvorwärmung f, Gemischanwärmung f
mixture preparation Gemischaufbereitung f,
Kraftstoffaufbereitung f,
mixture ratio Mischungsverhältnis n
mixture strength Mischungsverhältnis n
starting mixture Anlaßgemisch n
model Muster n, Modell n, Typ m
model range Modellreihe f
moment Moment n,m
moment of inertia Trägheitsmoment n
ignition moment Zündzeitpunkt m
monocoque construction Schalenbauweise f
motion Bewegung f
back motion roller Rücklaufrolle f
free motion Spiel n
reciprocating motion Hin- und Herbewegung f
motor Motor m
motor bicycle Moped n, Motorfahrrad n
motor brake Motorbremse f
motor car Personenkraftwagen m, Pkw m
motor drive Motorantrieb m
motor home Wohnmobil n
motor hood Motorhaube f
motor octane number Motor-Oktanzahl f
motor output Motorleistung f
motor school Fahrschule f
motor truck (lorry) Lastkraftwagen m
motor vehicle Kraftfahrzeug n
motor vehicle engineeering Kraftfahrzeugtechnik f
motor vehicle industry Automobilindustrie f
alternating-current motor Wechselstrommotor m
direct-current motor Gleichstrommotor m
fan motor Lüftermotor m
starter motor Anlaßmotor m
wiper motor Scheibenwischermotor m
motorcycle Motorrad n
mount/to montieren, anbringen
mounting Fassung f, Aufhängung f, Montage f
mounting clip Befestigungsschelle f

mounting device   Montagevorrichtung f, Aufziehvorrichtung f
mounting ga[u]ge   Montagelehre f
mounting lever   Montagehebel m
glass mounting   Glasfassung f
mud   Schlamm m, Matsch m
mud flap   Schmutzfänger m
battery mud   Batterieschlamm m
mudguard   Kotflügel m
muffler   Auspufftopf m, Schalldämpfer m
muffler cut-out   Auspuffklappe f
exhaust muffler   Auspuffschalldämpfer m, Auspufftopf m
multi   Mehrfach- *(s.a. multiple)*
multi-axle vehicle   Mehrachsfahrzeug n
multi-barrel carburetor   Mehrfachvergaser m,
Mehrstufenvergaser m, Vielstufenvergaser m, Stufenvergaser m,
Registervergaser m
multi-carburetor system   Mehrvergaseranlage f,
Mehrfachvergaseranlage f, ”Vergaserbatterie” f
multi-circuit switch   Serienschalter m
multi-fuel engine   Vielstoffmotor m, Mehrstoffmotor m
multi-fuel operation   Vielstoffbetrieb m
multi-jet carburetor   Mehrdüsenvergaser m
multi-plate clutch   Lamellenkupplung f, Mehrscheibenkupplung f
multi-point bearing   Mehrpunktlagerung f
multi-speed blower (fan)   mehrstufiges Gebläse n, mehrstufiger
Ventilator m
multi-speed gearbox   Mehrganggetriebe n, Vielganggetriebe n
multi-speed transmission   Mehrganggetriebe n,
Vielganggetriebe n
multi-stage carburetor   Mehrstufenvergaser m,
Vielstufenvergaser m, Stufenvergaser m, Registervergaser m,
Mehrfachvergaser m
multi-step reduction gear   Stufengetriebe n
multiplate   Mehrscheiben-
dry multiplate clutch   Mehrscheibentrockenkupplung f
multiple   Mehrfach-
multiple carburetor   Mehrfachvergaser m, Mehrstufenvergaser m,
Vielstufenvergaser m, Stufenvergaser m, Registervergaser m
multiple carburetor system   Mehrvergaseranlage f,

Mehrfachvergaseranlage f, "Vergaserbatterie" f
multiple circuit brake system Mehrkreisbremsanlage f
multiple hole nozzle Mehrlochdüse f, Viellochdüse f
multiple injection Mehrfacheinspritzung f
multiple jet carburetor Mehrdüsenvergaser m
multiplug Mehrfachstecker m
multiway valve Mehrwegventil n
mushroom valve Tellerventil n
narrow V-belt Schmalkeilriemen m
nationality plate Nationalitätszeichen n
naturally aspirated engine Saugmotor m
neck Stutzen m, Aussatz m
  neck journal bearing Halslager n
  neck of a bearing Lagerhals m
  filler neck Einfüllstutzen m
  oil filler neck Öleinfüllstutzen m
needle Nadel f
  needle bearing Nadellager n
  needle cage Nadelkäfig m
  needle control Nadelregelung f *(Nadelventil)*
  needle valve Nadelventil n
  float needle Schwimmernadel f
  float-needle valve Schwimmernadelventil n
  full-load needle Vollastnadel f
negative negativ
  negative camber negativer Sturz m
  negative caster Vorlauf m
  negative pole Minuspol m
neutral position Leerlaufstellung f, Neutralstellung f *(Getriebe)*
nipple Nippel m
no-load stroke Leer*l*aufhub m
nominal Nenn-
  nominal current Nennstrom m
  nominal dimension Nennmaß n
  nominal speed Nenndrehzahl f
non-misting screen Klarsichtscheibe f
non-return valve Rückströmventil n, Rückschlagventil n
non-skid tire Haftreifen m
non-slip rutschfest

non-sparking funkenfrei
normal speed Betriebsdrehzahl f, Normalgeschwindigkeit f
nose Vorbau m, Fahrzeugfront f, "Schnauze" f *(Karosserie)*
notch Raste f
notchback Stufenheck n
notched pin Kerbstift m
nozzle Düse f, Einspritzdüse f
  nozzle arrangement Düsenanordnung f
  nozzle body Düsenkörper m
  nozzle bore Düsenbohrung f
  nozzle holder Düsenhalter m, Düstenstock m
  nozzle hole Düsenbohrung f, Düsenloch f
  nozzle insertion Düseneinsatz m
  nozzle needle Düsennadel f
  nozzle opening Düsenöffnung f
  nozzle plunger Düsennadel f
  nozzle support Düsenhalter m
  nozzle tip Düsenkopf m, Düsenspitze f
  injection nozzle Einspritzdüse f
  spray nozzle Vergaserdüse f, Einspritzdüse f
  starting nozzle Starterdüse f
  suction nozzle Saugdüse f
number Zahl f, Nummer f; Kennzeichen n
  number of revolutions Umdrehungszahl f
  number plate *(polizeiliches)* Kennzeichen n, Kennzeichenschild n
  number-plate lamp Nummernschildleuchte f, Kennzeichenleuchte f
nut Mutter f
  adjusting nut Einstellmutter f , Stellmutter f
  cap nut Kappenmutter f, Hutmutter f
  castle nut Kronenmutter f
  connecting nut Anschlußmutter f
  counter nut Gegenmutter f
  discharge nut Ablaßmutter f
  fastening nut Befestigungsmutter f
  hexagon nut Sechskantmutter f
  knurled nut Rändelmutter f
  lock nut Gegenmutter f, Kontermutter f
  retaining nut Befestigungsmutter f

safety nut   Sicherungsmutter f
square nut   Vierkantmutter f
union nut   Anschlußmutter f
wheel nut spanner   Radmutterschlüssel m
wing nut   Flügelmutter f
oblique ball bearing   Schrägkugellager n
occasional seat   Notsitz m
offset   Versatz m
off-the-road   geländegängig, Gelände-
oil   Öl n
oil baffle   Schwallblech n in der Ölwanne, Ölschwallblech n,
Ölfangblech n, Ölleitrippe f, Ölablenkblech n, Ölleitblech n
oil bath air filter   Ölbadluftfilter n,m
oil catch ring   Ölauffangring m
oil catcher   Ölfangblech n
oil change   Ölwechsel m
oil change interval   Ölwechselintervall n
oil circulation   Ölkreislauf m, Ölumlauf m
oil container   Ölfangbecken n, Ölbehälter m
oil control ring   Ölabstreifring m
oil cooler   Schmierölkühler m, Ölkühler m
oil deflector sheet   Ölabstreifblech n
oil dip stick   Ölmeßstab m
oil drain pipe   Ölablaßstutzen m
oil drain plug   Ölablaßschraube f
oil drain tap   Ölablaßhahn m
oil drain valve   Ölablaßventil n
oil duct   Ölleitung f, Ölkanal m, Ölbohrung f
oil feed   Ölversorgung f, Ölzufuhr f
oil feed line   Ölversorgungsleitung f, Ölzufuhrleitung f
oil filler neck [pipe]   Öleinfüllstutzen m
oil filling screw   Öleinfüllschraube f
oil filling tube   Öleinfüllrohr n
oil film   Schmierfilm m
oil filter   Schmierölfilter n,m, Ölfilter n,m
oil filter cartridge   Ölfilterpatrone f
oil filter insert   Ölfiltereinsatz m
oil gallery   Ölkanal m
oil groove   Schmiernut f, Ölrinne f

oil jet Öldüse f
oil leakage Ölundichtheit f, Ölverlust m
oil level ga[u]ge Ölstandmeßgerät n
oil level indicator Ölstandanzeiger m
oil line Ölleitung f
oil pan Ölwanne f, Kurbelwanne f
oil pipe Ölleitung f, Ölstutzen m
oil plug Ölschraube f
oil pressure Öldruck m
oil pressure cylinder Öldruckzylinder m
oil pressure ga[u]ge Öldruckmeßgerät n
oil-proof ölbeständig
oil pump connection Ölpumpenanschluß m
oil pump drive Ölpumpenantrieb m
oil pump housing Ölpumpengehäuse n
oil pump support Ölpumpenlagerung f
oil retainer Ölspritzblech n, Ölschutzblech n, Ölfangblech n
oil retainer ring Ölspritzring m
oil return line Ölrücklaufleitung f, Ölrückführungsleitung f,
Ölleitung f vom Ölkühler
oil return pipe Ölrücklaufrohr n
oil ring Schmierring m, Ölabstreifring m
oil scoop Ölschöpfer m
oil scraper ring Ölabstreifring m
oil screen Ölsieb n
oil separator Ölabscheider m
oil spray nozzle Ölspritzdüse f
oil sump Ölsumpf m
oil sump gasket Ölwannendichtung f
oil supply Ölversorgung f, Ölzufuhr f
oil supply bore Ölbohrung f
oil tank Schmierölbehälter m, Öltank m
oil trap Ölabscheider m, Entöler m
oil vapo[u]r Öldampf m, Öldunst m
oil-wetted air cleaner Naßluftfilter n,m
diesel oil Dieselöl n
heavy-oil carburetor Schwerölvergaser m
leakage oil line Leckölleitung f
lubricating oil Schmieröl n

running-in oil Einlauföl n
rustproof oil Korrosionsschutzöl n
transmission oil Getriebeöl n
oiler Öler m, Schmiervorrichtung f, Schmiernippel m
one Eins f, Ein-, Einzel-
  one-piece propeller shaft einteilige Gelenkwelle f, einteilige Kardanwelle f
  one-piece rim einteilige Felge f
  one-speed blower einstufiges Gebläse n, einstufiger Ventilator m
  one-spoke steering wheel Einspeichenlenkrad n
  one-way clutch Freilaufkupplung f, Freilauf m
  one-way valve Einwegventil n
opening Öffnung f, Aussparung f
  opening angle Öffnungswinkel m
  opening pressure Abspritzdruck m *(Einspritzung)*
  air bleed opening Nebenlufteinlaß m
  nozzle opening Düsenöffnung f
operate/to betätigen
operated betätigt
  cable-operated seilzugbetätigt
  centrifugally operated timing controller Fliehkraftverteilerversteller m
  centrifugally operated timing device Fliehkraft[zündzeitpunkt]versteller m
  crank-operated window Kurbelfenster n
  exhaust-operated air heating Abgasheizung f
  foot-operated dimming switch Fußabblendschalter m
  foot-operated switch Fußschalter m
  pedal-operated dip switch Abblendfußschalter m
operating Betriebs-, Arbeits-, Funktions-
  operating current Betriebsstrom m
  operating cycle Arbeitsspiel n *(Motor)*
  operating cylinder Betätigungszylinder m
  operating instruction Bedienungsanleitung f
  operating lever Betätigungshebel m
  operating pressure Betriebsdruck m
  operating speed Betriebsdrehzahl f
  operating temperature Betriebstemperatur f
  operating test Funktionsprüfung f

operating voltage Betriebsspannung f
door-operating cylinder Türbetätigungszylinder m
operation Betätigung f
  hand (manual) operation Handbetätigung f
opposed entgegengesetzt, Gegen-
  opposed current Gegenstrom m
  opposed cylinder [piston] engine Boxermotor m,
  Gegenkolbenmotor m
optional equipment Sonderausstattung f, Sonderzubehör n
options Sonderausstattung f, Sonderzubehör n, Extras f
order Ordnung f, Klasse f, Sorte f, Reihenfolge f
  firing (ignition) order Zündfolge f
orifice Mündung f, Öffnung f, Düse f, Düsenbohrung f *(Vergaser)*
original Original-, Erst-
  original equipment Erstausstattung f
  original ring Originalring m
ornamental Zier-
  ornamental fittings Zierbeschläge f
  ornamental frame Zierrahmen m
oscillation Schwingung f
otto engine Ottomotor m
outer Außen-
  outer diameter Außendurchmesser m
  outer skin Mantel m *(Karosserie)*
  outer (skin) panel Außenblech n *(Karosserie)*
outlet Ausgang m, Auslaß m, Steckanschluß m
  outlet connection Auslaufstutzen m
  outlet pipe Ablaßrohr n, Austrittstutzen m, Auslaufstutzen m
  outlet shaft Austrittswelle f
  outlet temperature Abgastemperatur f
  outlet valve Ablaßventil n
output Leistung f, Arbeitsabgabe f, Ausgangsleistung f
  output cylinder Nehmerzylinder m
  output end Abtriebseite f, Motor-Abtriebseite f
  output end cover Seitenteil-Abtriebseite f
  output flange Abtriebflansch m
  output oil-pump Sekundärölpumpe f, Sekundärpumpe f
  output per cylinder Zylinderleistung f
  output rotor Turbinenrad n

output shaft Abtriebswelle f, Ausgangswelle f
output torque Abtriebsdrehmoment n, Ausgangsdrehmoment n
power output Leistungsabgabe f
rated output Nennleistung f
starting output Anlaßleistung f
outrigger Ausleger m
outside außerhalb, Außen-
  outside air Außenluft f
  outside diameter Außendurchmesser m
  outside mirror Außenspiegel m
  outside rear-view mirror Außenrückspiegel m
oval head screw Linsenschraube f
overall Gesamt-
  overall efficiency Gesamtwirkungsgrad m
  overall ratio Gesamtübersetzung f
overdrive Schnellgang m, Schongang m, Spargang m, Ferngang m
  overdrive unit Schnellganggetriebe n
overflow Überlauf m, Übertröm-
  overflow conduit Überlaufleitung f
  overflow cup Überlaufbecher m
  overflow duct Überströmkanal m
  overflow jet Überlaufdüse f
  overflow tank Überlaufbehälter m
  overflow tube Überlaufrohr n
  overflow valve Überströmventil n
overhauling Überholung f, Durchsicht f
overhead valve hängendes (obengesteuertes) Ventil n
overheating Überhitzung f
overload Überlast f
  overload capacity Überlastbarkeit f
  overload protection Überlastschutz m
  overload valve Überdruckventil n
override clutch gear change Überholklauenschaltung f
overrun brake Auflaufbremse f
overrun conditions Schubbetrieb m *(Motor)*
overrunning clutch Freilaufkupplung f, Überholkupplung f
overspeed governor Drehzahlbegrenzer m
overspeeding Überschreitung f der zulässigen Höchstgeschwindigkeit
oversteer/to übersteuern

owner's manual   Kundendienstheft n
oxide magnet   Oxidmagnet m
packing   Dichtung f, Packung f
  packing ring   Abdichtring m, Dichtungsring m, Dichtring m
pad   Unterlage f, Bremsbelag n, Belag m
  seat pad   Sitzkissen n
  spring pad   Federunterlage f, Federauflage f
paint   Lack m
  paint coating   Lackschicht f
painting box   Spritzkabine f
paintshop   Lackiererei f
paintwork   Lackierung f
pan   Wanne f, Schale f
  oil pan   Ölfänger m, Ölwanne f
pane   Glasscheibe f, Scheibe f
  pane framing   Scheibenumrahmung f
  door pane   Türscheibe f
  light exit pane   Lichtaustrittscheibe f
panel   Karosserieblech n, Abdeckblech n, Blende f
  panel lamp   Armaturenleuchte f
  panel lighting   Armaturenbrettbeleuchtung f
  back panel   Heck n *(Karosserie)*
  door outside panel   Türaußenblech n
  instrument panel   Armaturenbrett n, Instrumententafel f
  rear panel   Heck n *(der Karosserie)*
panel[l]ing   Verkleidung f, Seitenverkleidung f
Panhard rod   Panhard-Stab m, Querlenker m
parabolic reflector   Parabolspiegel m
parcel shelf   Ablagefach n
park/to   parken
parking   Parken n
  parking brake   Feststellbremse f
  parking lamp   Parkleuchte f
  parking light   Parklicht n, Standlicht n
  parking position   Parkstellung f
part   Teil n, Teilstück n
  part load   Teillast f
  part range   Teillastbereich m
  replacement part   Ersatzteil n

   counter part Gegenstück n
   spare part Ersatzteil n
partial teilweise, Teil-
   partial coverage insurance Teilkaskoversicherung f
   partial-flow filter Nebenstromfilter n,m
   partial load Teillast f
   partial load range Teillastbereich m
   partial-load valve Teillastventil n
particulate emission Feststoffemission f, Teilchenemission f
partition Trennung f, Teilung f, Trennwand f
   partition panel (wall) Trennwand f
parts list Stückliste f
parts set Teilesatz m
pass/to überholen
passenger Fahrzeuginsasse m, Beifahrer m, Fahrgast m
   passenger car Personenkraftwagen m, PKW m
   passenger compartment Fahrgastraum m, Innenraum m
passing Überholen n
   passing beam Abblendlicht n
   passing lane Überholspur f
   passing light Abblendlicht n
path Weg m
   brake-path length Bremsweg m
pattern Muster n; Profil n *(Reifen)*
   gear-shift pattern Ganganordnung f
pawl Klinke f, Sperrklinke f
   locking pawl Schließraste f
   ratchet pawl Sperrklinke f
   safety pawl Sicherungsklinke f
payload Nutzlast f, Zuladung f
peak Spitze f, Höchstwert m, Maximum n
   peak current Spitzenstrom m
   peak power Höchstleistung f, maximale Leistung f
   peak pressure Spitzendruck m, Maximaldruck m
   load peak Belastungsspitze f
pedal Fußhebel m, Fußpedal n, Pedal n
   pedal application Fußhebelbetätigung f
   pedal bracket Fußhebellagerbock m
   pedal brake Fußbremse f

pedal clearance Pedalspiel n
pedal gear Pedalwerk n, Fußhebelwerk n
pedal lining Trittplattenüberzug m, Pedalauflage f
pedal-operated dip switch Abblendfußschalter m
pedal operation Fußhebelbetätigung f
pedal play Pedalspiel n
pedal stop Pedalanschlag m
pedal travel Pedalweg m
accelerator pedal Gaspedal n
clutch pedal Kupplungspedal n
performance Ausführung f, Leistung f, Verhalten n
performance data Leistungswerte mpl
period Periode f, Zeitdauer f
period of vibration Schwingungsdauer f
injection period Einspritzdauer f, Öffnungsdauer f, Ventilanhebezeit f
running-in period Einlaufzeit f
petroil Zweitaktgemisch n
petroil lubrication Gemischschmierung f, Zweitaktschmierung f
petrol Benzin n
petrol consumption Benzinverbrauch m
petrol-dispensing pump Benzin-Zapfsäule f, Zapfsäule f
petrol engine Benzinmotor m
petrol ga[u]ge Benzinuhr f, Benzinstandsmeßgerät n
petrol injection Benzineinspritzung f
petrol jet Benzindüse f, Kraftstoffdüse f
petrol tank Benzintank m
petrol warning light Benzinkontrolleuchte f
petroleum Erdöl n, Rohöl n, Petroleum n
pick-up/to 1. jemanden mit dem PKW abholen; 2. zügig beschleunigen
pick-up 1. Kleinlastwagen m, leichter LKW m; 2. Anfahrbeschleunigung f; 3. Ansprechen n *(Vergaser)*
pick-up behavio[u]r Fahrverhalten n, Ansprechverhalten n
pick-up van Lieferwagen m
piece Stück n
connecting piece Anschlußteil n, Anschlußstück n
flexible tube piece Schlauchstück n
fusible piece Schmelzeinsatz m

sliding piece   Gleitstück n
stop piece   Anschlagstück n
threaded piece   Gewindestück n
tube piece   Rohrstück n
pillar   Säule f, Pfosten m
  A-pillar   vordere Säule f, A-Säule f
  B-pillar   Mittelsäule f, B-Säule f
  C-pillar   hintere Säule f, C-Säule f
pilot   Vor-, Start-, Leer-
  pilot air   Leerlaufluft f
  pilot air jet   Leerlaufluftdüse f, Leerlaufbohrung f
  pilot air screw   Leerlaufluftschraube f,
  Leerlaufluft-Einstellschraube f
  pilot-air system   Leerlaufluftsystem n
  pilot injection   Voreinspritzung f
  pilot jet   Vor[einspritz]düse f, Starterdüse f, Starterbohrung f
  pilot lamp   Kontrollampe f, Anzeigelampe f
  pilot light   Kontrollicht n, Anzeigelicht n
  pilot run   Nullserie f, Vorserie f
pin   Stift m, Nadel f, Bolzen m, Kolbenbolzen m
  ball pin   Kugelbolzen m
  bearing pin   Lagerstift m, Lagerzapfen m
  code pin   Codierstift m
  contact pin   Kontaktstift m
  crank pin   Kurbelzapfen m
  cylindrical pin   Zylinderstift m
  disengaging pin   Ausrückstift m
  driving pin   Antriebsbolzen m, Mitnehmerbolzen m
  follower pin   Mitnehmerstift m, Mitnehmerzapfen m
  guide pin   Führungsstift m, Führungszapfen m
  king-pin angle   Radspreizungswinkel m
  king-pin inclination   Spreizung f
  king-pin steering [assembly]   Achsschenkellenkung f
  notched pin   Kerbstift m
  pivot pin   Drehstift m
  split pin   Splint m
  straight pin   Zylinderstift m
  threaded pin   Gewindestift m
pinion   Ritzel n

pinion drive shaft   Ritzelwelle f, Ritzelantriebswelle f
pinion gear   Planetenrad n
pinion shaft   Ritzelwelle f
bevel pinion   Kegelrad n
differential bevel pinion   Ausgleichkegelrad n, kleines
Differentialantriebskegelrad n
driving pinion   Antriebsritzel n, Antriebszahnrad n
spur pinion   Stirnkegelrad n, Stirnrad n
starter pinion   Anlasserritzel n
pintle   Düsenzapfen m, Düsennadelzapfen m
pintle injector   Zapfendüse f
pintle nozzle   Zapfendüse f
pipe   Rohr n, Leitung f
pipe clamp (clip)   Befestigungsschelle f, Rohrschelle f
pipe connection   Rohranschluß m, Rohrverbindung f
pipe joint   Rohrdichtung f, Rohrverschraubung f
pipe line   Rohrleitung f
pipe socket (union)   Rohrstutzen m
air relief pipe   Entlüfterrohr n
collecting pipe   Sammelrohr n
connecting pipe   Anschlußrohr n
cooling pipe   Kühlleitung f
drain pipe   Ablaßstutzen m, Austrittsstutzen m
exhaust pipe   Auspuffrohr n, Auspuff m
feed pipe   Zuleitungsrohr n
flexible pipe   biegsames Rohr n
induction pipe   Ansaugrohr n, Einlaßrohr n
injection pipe   Einspritzleitung f
inlet pipe   Einlaufstutzen m
intake pipe   Ansaugrohr n
outlet pipe   Ablaßrohr n
pressure pipe   Druckleitung f, Druckstutzen m
return pipe   Rücklaufleitung f
suction pipe   Ansaugrohr n, Saugleitung f, Saugrohr n
water pipe   Wasserstutzen m
piston   Kolben m *(Motor)*; Bremskolben m
piston barrel   Kolbenkörper m
piston body   Kolbenmantel m
piston boss   Kolbenauge n, Kolbennabe f, Kolbenbolzenauge n,

Bolzenauge n, Kolbenbolzennabe f
piston-boss bushing Kolbenbuchse f
piston clearance Kolbenpassung f, Kolbenspiel n, Kolbenluft f
piston crown Kolbenboden m
piston dead center Kolbentotlage f
piston displacement Hubraum m
piston engine Kolbenmotor m, Hubkolbenmotor m
piston groove Kolbenringnut f, Ringnut f
piston guide Kolbenführung f
piston head Kolbenboden m
piston pin Kolbenbolzen m, Kolbenzapfen m
piston pin boss Kolbenbolzenauge n, Kolbenbolzennabe f
piston pin circlip Kolbenbolzensicherungsring m,
Kolbenbolzennabe f
piston play Kolbenspiel n, Kolbenluft f
piston pump Kolbenpumpe f
piston ring Kolbenring m, Spannring m
piston ring groove Kolbenringnut f
piston ring joint Kolbenringstoß m
piston rod Pleuelstange f, Pleuel n, Kolbenstange f
piston scavenging Kolbenspülung f
piston skirt Kolbenschaft m, Schaft m, Kolbenmantel m,
Kolbenhemd n
piston slack Kolbenschlag m
piston stroke Kolbenhub m, Kolbentakt m
piston top Kolbenboden m
piston top land Feuersteg m, Bodensteg m, oberster Kolbensteg n
piston travel Kolbenweg m, Kolbenhub m
control piston Steuerkolben m
distributor piston Verteilerkolben m
pit 1. Reparaturgrube f, Box f; 2. Laufflächenmulde f *(Reifen)*
pitch Teilung f, Zahnteilung f *(Zahnrad)*; 2. Nicken n
(Fahrzeugbewegung)
pitch circle Teilkreis m
pitch diameter Teilkreisdurchmesser m
coarse pitch thread Steilgewinde n
winding pitch Wicklungsschritt m
pitman arm Lenkstockhebel m, Lenkhebel m
pivot Drehbolzen m, Drehzapfen m , Zapfen m

plate   **143**

pivot bearing   Drehlager n
pivot pin   Drehstift m, Lagerbolzen m
pivot screw   Zapfenschraube f
  threaded pivot   Gewindezapfen m
plain **combustion chamber**   offener Brennraum m
  (Direkteinspritzmotor)
planet wheel   Planetenrad n
planetary   Planeten-
  planetary gear   Planetengetriebe n
  planetary wheel   Planetenrad n, Umlaufrad n
plant   Werk n, Fabrik f, Anlage f
  air-conditioning plant   Klimaanlage f
  electric light plant   Lichtanlage f
  washing plant   Waschanlage f
plastic   Kunststoff m, Plastik-
  plastic body   Kunststoffkarosserie f, Plastikkarosserie f
  plastic coating   Kunststoffbeschichtung f, Plastikbeschichtung f
  plastic foil   Kunststoffolie f, Plastikfolie f
  plastic fuel tank   Kraftstoffbehälter m aus Kunststoff,
  Kraftstofftank m aus Kunststoff, Kunststofftank m, Plastiktank m
  plastic insert   Kunststoffeinsatz m
plate   Schild n, Blech n, Scheibe f, Platte f
  baffle plate   Prallblech n
  base plate   Grundplatte f
  clutch plate   Kupplungsscheibe f
  clutch pressure plate   Kupplungsdruckplatte f
  connecting plate   Anschlußplatte f
  contact breaker plate   Unterbrecherplatte f
  contact plate   Kontaktplatte f, Aufschlagplatte f
  coupling plate   Kupplungsplatte f
  cover plate   Abdeckplatte f, Deckblech n, Deckplatte f
  covering plate   Abdeckplatte f, Deckblech n, Deckplatte f
  deflector plate   Spritzblech n
  dry-plate clutch   Trocken[scheiben]kupplung f
  identification plate   Typenschild n
  joint plate   Anschlußblech n
  licence plate   Nummernschild n
  licence plate lamp   Kennzeichenleuchte f, Nummernschildleuchte f
  multi-plate clutch   Lamellenkupplung f

multi-plate friction clutch   Lamellenrutschkupplung f
number plate   polizeiliches Kennzeichen n
pressure plate   Druckplatte f *(z.B. zur Kupplung)*
protective plate   Schutzplatte f
screening plate   Abschirmblech n
spring plate   Federteller m
stiffening plate   Verstärkungsblech n
platform   Ladefläche f, Pritsche f; Werkstattbühne f, Bühne f
platform gate   Bordwand f *(Lastkraftwagen)*
platform truck   Pritschenwagen m, Pritschen-LKW m
play   Spiel n, Luft f
bearing play   Lagerluft f
pliers   Zange f
combination pliers   Kombizange f
crimp pliers   Quetschzange f, Crimpzange f
plug   Verschlußstopfen m, Stöpsel m, Stecker m, Zündkerze f
plug-and-socket connection   Steckverbindung f
plug housing   Steckergehäuse n
plug socket   Steckdose f, Hülse f, Zündkerzenstecker m
plug spanner   Zündkerzenschlüssel m
battery cell plug   Zellenstopfen m
cold plug   Zündkerze f mit hohem Wärmewert
coupling plug   Kupplungsstecker m
drain plug   Ablaßschraube f
filler plug   Einfüllschraube f
fuse plug   Schmelzeinsatz m, Sicherungseinsatz m
glow (heater) plug   Glühkerze f
heater plug resistor   Glühkerzenwiderstand m
heater plug starting switch   Glühanlaßschalter m
heating plug   Glühkerze f
hot plug   Zündkerze f mit niedrigem Wärmewert
ignition cable plug   Zündleitungsstecker m
oil plug   Ölschraube f
radio-shielding plug   Entstörstecker m
screw plug   Verschlußstopfen m
spark plug   Zündkerze f
spark plug air gap   Elektrodenabstand m
spark plug socket   Zündkerzenstecker m
spark plug wire   Zündkabel n

   suppressing plug Entstörstecker m
   vent plug Entlüfterpfropfen m
plunger Kolben m, Bremskolben m
   plunger rod Kolbenstange f
   plunger stroke Kolbenhub m
   delivery plunger Förderkolben m
ply Lage f, Kordlage f *(Reifen)*
   cross-ply tire Diagonalreifen m
   radial-ply tire Radialreifen m
pneumatic pneumatisch, Druckluft-
   pneumatic brake Druckluftbremse f
   pneumatic brake system Druckluftbremsanlage f
   pneumatic gear change Druckluftschaltung f
   pneumatic gear shift Druckluftschaltung f
   pneumatic tire Luftreifen m
point Spitze f, Punkt m
   contact breaker point Unterbrecherkontakt m
   jacking point Ansatzstelle f für Wagenheber
   multi-point bearing Mehrpunktlagerung f
pointer Zeiger m
pole Pol m, Pfosten m
   pole casing [housing] Polgehäuse n
   connecting pole Anschlußpol m
   negative pole Minuspol m
   positive pole Pluspol m
   single-pole einpolig
   two-pole zweipolig
pollutant emission Schadstoffausstoß m
pollution Verschmutzung f, Schadstoffausstoß m
pool Fuhrpark m
port Kanal m, Steueröffnung f, Steuerschlitz m
   port injection Saugkanaleinspritzung f
   port liner Kanalbuchse f
   port timing Steuerzeiten fpl
   induction port Ansaugkanal m, Einlaßkanal m
   injection port Einspritzöffnung f
   suction port Ansaugschlitz m
position Stellung f, Lage f
   position lamp Positionslampe f, Begrenzungsleuchte f

parking position   Parkstellung f
positive   positiv
  positive camber   positiver Sturz *(Rad)*
  positive caster   Nachlauf m
  positive plate   Plusplatte f
  positive pole   Pluspol m
  positive terminal   Plusklemme f, Pluspol m
post   Pfosten m, Säule f *(Karosserie)*
  A-post   A-Säule f, vordere Säule f
  B-post   B-Säule f, Mittelsäule f
  C-post   C-Säule f, hintere Säule f
  door post   Türpfosten m
power   Leistung f, Kraft f, Stärke f
  power assistance   Verstärker m
  power-assisted   mit Bremshilfe ausgerüstet, servounterstützt
  power-assisted brake system   Hilfskraftbremsanlage f,
Bremsanlage f mit Servounterstützung
  power-assisted steering   Servolenkung f
  power booster   Bremskraftverstärker m,
Bremskraftunterstützungsgerät n
  power brake   Servobremse f, Bremsanlage f mit Servounterstützung
  power brake system   Fremdkraftbremsanlage f
  power consumption   Leistungsaufnahme f, Leistungsbedarf m,
Leistungsverbrauch m
  power cycle   Arbeitstakt m
  power cylinder   Arbeitszylinder m, Servozylinder m *(Servolenkung)*
  power input   Leistungsaufnahme f
  power loss   Leistungsverlust m, Leistungsabfall m
  power output   Motorleistung f, Leistungsabgabe f, abgegebene
Leistung f
  power per unit of displacement   Hubraumleistung f,
Literleistung f
  power plant   Triebwerk n
  power steering   Servolenkung f, Hilfkraftlenkanlage f,
Fremdkraftlenkanlage f
  power steering gear   Servolenkungsgetriebe n
  power steering mechanism   Servolenkung f, Servolenkgetriebe n
  power steering pump   Lenkhelfpumpe f, Lenkhilfspumpe f
  power steering system   Hilfskraftlenkanlage f,

Fremdkraftlenkanlage f, Servolenkung f
power stroke Arbeitshub m, Arbeitstakt m
power train Kraftübertragung f
power transmission Kraftübertragung f
main power cut-off Batteriehauptschalter m
starting power Startleistung f
preassembly Vormontage f
preexpansion chamber Auspuffvorschalldämpfer m
prechamber Vorkammer f
precombustion chamber Vorverbrennungskammer f, Vorkammer f
prefilter Vorfilter n,m
preheating flap (valve) Vorwärmklappe f
preignition Glühzündung f, Frühzündung f, Vorzündung f
preliminary Vor-
  preliminary air tank Vorluftbehälter m
  preliminary filter Vorfilter n,m
premature ignition Frühzündung f
preselection gear shift Vorwählschaltung f
preselector Vorwählschalter m
  preselector lever Vorwählhebel m
  preselector transmission Vorwählgetriebe n
preservative agent Konservierungsmittel n
presilencer Vorschalldämpfer m
pressure Druck n, Druckkraft f
  pressure accumulator Druckspeicher m
  pressure angle Eingriffswinkel m
  pressure casting Druckguß m
  pressure control valve Drucksteuerventil n
  pressure cylinder Druckzylinder m
  pressure diecasting Druckguß m
  pressure differential indicator Differenzdruckanzeiger m,
Bremsdruckdifferenzanzeiger m, Druckdifferenzanzeiger m
  pressure ga[u]ge Druckmeßgerät n, Manometer n
  pressure indicator Druckanzeigegerät n
  pressure limiting valve Druckbegrenzungsventil n
  pressure lubrication system Druckumlaufschmierung f,
Druckschmierung f
  pressure pipe Druckleitung f
  pressure plate Druckplatte f, Kupplungsdruckplatte f

pressure pump   Druckpumpe f
pressure reducing valve   Druckreduzierventil n,
Druckminderventil n
pressure regulating valve   Druckreglerventil n
pressure regulator   Druckregler m
pressure sensor   Druckfühler m, Drucksensor m
pressure switch   Druckschalter m
pressure valve   Druckventil n
pressure vessel   Druckbehälter m
pressure-wave supercharger   Druckwellenlader m,
Druckwellenkompensator m
adjusting pressure   Einstelldruck m
air-pressure ga[u]ge   Luftdruckprüfer m *(Reifen)*
contact pressure   Kontaktdruck m
discharge pressure   Abspritzdruck m
high-pressure hose   Hochdruckschlauch m
inflation pressure   Reifen[füll]druck m, Reifeninnendruck m
initial pressure   Anfangsdruck m
injection pressure   Einspritzdruck m, Öffnungsdruck m
low pressure   Unterdruck m
setting pressure   Einstelldruck m
tire-inflation pressure   Reifeninnendruck m
working pressure   Arbeitsdruck m
pressurized carburetor   Druckvergaser m
primary   primär, elementar
primary barrel   erste Stufe f
primary circuit   Primärstromkreis m, Primärkreis m
primary coil   Primärspule f
primary filter   Vorfilter n,m
primary fuel filter   Kraftstoffvorfilter n,m
primary stage   erste Stufe f
primary winding   Primärwicklung f
primer   Grundlack m, Grundierung f
printed circuit   gedruckte Schaltung f
priority   Vorfahrt f, Vorrang m, Priorität f
priority vehicle light   Rundum-Kennleuchte f für vorrangige
Fahrzeuge
propeller shaft   Gelenkwelle f, Kardanwelle f
propeller shaft tunnel   Kardantunnel m, Mitteltunnel m

probe   Sonde f, Meßfühler m
procedure   Arbeitsweise f
proceed/to   weiterfahren
proceed ahead/to   geradeaus weiterfahren
process   Prozeß m, Verfahren n, Vorgang m
  starting process   Anlaßvorgang m
  working process   Arbeitsgang m
production car   Serienwagen m
production part   Serienteil n
profile   Profil n
  profile ledge   Profilleiste f
  profile rubber   Profilgummi m, Formgummi m
  profile tube   Profilrohr n
proof   widerstandsfähig, beständig, fest
  acid-proof   säurebeständig, säurefest
  acid-proof grease   Säureschutzfett n
  corrosion-proof   rostbeständig, rostgeschützt, korrosionsgeschützt
  door proof catch   Türsicherung f
  dust-proof   staubdicht
  short-circuit proof   kurzschlußsicher
  tropic-proof .tropenfest
property   Eigenschaft f
  cross-country property   Geländegängigkeit f
  property damage insurance   Haftpflichtversicherung f
propulsion   Antrieb m
  propulsion power   Antriebskraft f, Antriebsleistung f
  propulsion shaft   Antriebswelle f
protect/to   schützen, sichern
  protect by fuse/to   absichern *(elektrischer Stromkreis)*
protecting   Schutz-
  protecting cap   Schutzkappe f
  protecting cover   Schutzdeckel m
protection   Schutz m
  corrosion protection   Korrosionsschutz m
  starter protection   Anlasserschutz m
  overload protection   Überlastschutz m
protective   schützend, Schutz-
  protective cap   Schutzkappe f
  protective cover   Schutzdeckel m

protective covering Schutzhülle f, Isolierschlauch m *(Kabel)*
protective mo[u]lding [strip] Schutzleiste f, Scheuerleiste f
protective switch Schutzschalter m
prototype Prototyp m
protruding bumper vorstehender Stoßfänger m
pull Zug m
  pull-on brake Zughandbremse f
  pull switch Zugschalter m
pulldown Kaltstartöffnung f *(Starterklappe am Vergaser)*
pulley Riemenscheibe f, Rolle f
  belt pulley Riemenscheibe f
  driven (driving) pulley Antriebsscheibe f
  fan pulley Ventilatorscheibe f
  guide pulley Leitblock m
  idler pulley Leerlaufscheibe f
  reversing pulley Umlenkrolle f
  tension pulley Spannrolle f
  V-belt pulley Keilriemenscheibe f
pulse generator Zündimpulsgeber m, Impulsgeber m
pump Pumpe f
  pump barrel Pumpenzylinder m
  pump connection Pumpenanschluß m
  pump cylinder Pumpenzylinder m
  pump delivery Pumpenförderleistung f
  pump diaphragm Pumpenmembran f
  pump element Pumpenelement n
  pump housing Pumpengehäuse n
  pump piston (plunger) Pumpenkolben m
  air pump Luftpumpe f
  anti-freeze pump Frostschutzpumpe f
  circulation pump Umwälzpumpe f
  feed pump Förderpumpe f, Kraftstoffpumpe f
  fuel injection pump Einspritzpumpe f
  fuel pump Kraftstoffpumpe f
  gear pump Zahnradpumpe f
  hand feed pump Handförderpumpe f
  high-pressure pump Hochdruckpumpe f
  oil pump Schmierölpumpe f, Ölpumpe f
  piston pump Kolbenpumpe f

pressure pump Druckpumpe f
servo pump Servopumpe f
suction pump Absaugpumpe f
vacuum fuel pump Unterdurckförderpumpe f
vacuum pump Vakuumpumpe f
water pump Wasserpumpe f
puncture Reifenpanne f
purge line Spülleitung f
push Schub m, Stoß m, Schieben n, Stoßen n
  push button Druckknopf m, Drucktaste f
  push button switch Druckknopfschalter m, Druckschalter m
  push rod Schubstange f, Stoßstange f, Stößelstange f
  horn push button Horndruckknopf m
  horn push ring Horndruckring m
  light push button Leuchtdrucktaster m, Leuchtdrücktaste f
  light push switch Lichtschubschalter m
  pump push rod Pumpenstößel m
  screw-push starter Schubschraubtriebanlasser m
  signal push button Signaldruckknopf m
  starter push-button Anlaßschalter m, Anlasserdruckknopf m
  valve push rod Ventilstoßstange f
quality Qualität f
  quality control Qualitätskontrolle f, Gütekontrolle f,
  Gütesicherung f, Serienüberwachung f
quarter vent (window) Ausstellfenster n
quench area (zone) Quetschkante f, Quetschfläche f
quick charger Schnelladegerät n *(für Batterie)*
race/to rennen, Rennen fahren; rasen
race Rennen n; Laufbahn f
  ball race Laufring m
racing Renn-
  racing car Rennwagen m, Rennfahrzeug n
  racing circuit Rennstrecke f
  racing engine Rennmotor m
  racing track Rennstrecke f
rack Zahnstange f, Gestell n, Gepäckträger m
  rack-and-pinion steering Zahnstangenlenkung f
  rack-and-pinion steering gear Zahnstangenlenkgetriebe n
  roof rack Dachgepäckträger m

radar trap Radarfalle f
radial radial, Radial-
  radial clearance Radialspiel n
  radial compressor Radialgebläse n, Radialverdichter m
  radial engine Sternmotor m
  radial joint Radialdichtung f
  radial play Radialspiel n
  radial-ply carcass Radialkarkasse f
  radial-ply tire Radialreifen m, Gürtelreifen m
  radial run-out Höhenschlag m, Radialschlag m,
  Rundlaufabweichung f
  radial tire Radialreifen m, Gürtelreifen m
  radial tire run-out Höhenschlag m des Reifens,
  Reifenhöhenschlag m, Radialschlag m des Reifens,
  Rundlaufabweichung f des Reifens
radiation Strahlung f
radiator Kühler m, Wasserkühler m
  radiator block Kühlerblock m
  radiator cap Kühlerdeckel m, Einfüllverschluß m
  radiator core Kühlerblock m
  radiator flap Kühlerklappe f
  radiator frame Kühlerrahmen m, Kühlergehäuse n
  radiator grille Kühlergrill m
  radiator shell Kühlergehäuse n, Kühlerrahmen m
  radiator tank Wasserkasten m
  tubular radiator Röhrenkühler m
  cellular radiator Lamellenkühler m
radio Radio n
  radio communication system Sprechfunkanlage f,
  Funksprechanlage f
  radio set Autoradio n
  radio-shielding cable Entstörkabel n
  radio-shielding capacitor Entstörkondensator m
  radio-shielding plug Entstörstecker m
  radio telephone Funktelefon n, Autotelefon n
radius Radius m
  radius arm (link) Schwingarm m, Schrägschwinge f,
  schrägliegender Längslenker m, schrägliegende Längsabstützung f,
  Achsstrebe f, Schubstrebe f, Schubstange f, Zugstange f,

Schwingarm m, schrägliegender Längslenker m, schrägliegende
Längsabstützung f, Zugstange f, Schubstrebe f, Schubstange f
rail   Schiene f; Längsträger m, Längsholm m *(Rahmen)*
   contact rail   Kontaktschiene f
   guide rail   Führungsschiene f
   slide rail   Laufschiene f, Gleitschiene f
   stop rail   Anschlagschiene f
rain channel (groove, gutter)   Regenrinne f *(Karosserie)*
ram   Ramme f, Anschlag m, Stau-
   ram induction   Staudruckladung f
   ram pipe   Staurohr n *(Ansaugsystem)*
   ram pressure   Staudruck m
range   1. Bereich m; Entfernung f, Reichweite f; 2. Baureihe f
   measuring range   Meßbereich m
   range-change gearbox   Gruppengetriebe n,
   Mehrbereichsgetriebe n
   speed range   Drehzahlbereich m
   timing range   Verstellbereich m *(Zündung)*
ratchet   Ratsche f, Sperre f, Sperrklinke f
   ratchet brake   Ratschenbremse f *(Handbremse)*
   ratchet segment   Klinkensegment n
   ratchet wheel   Schaltrad n, Sperrad n
   gate roller ratchet   Gatterwalzenkappe f
rate of air flow   Luftdurchsatz m
rate of speed   Fahrgeschwindigkeit f
rated   Nenn-
   rated capacity   Nennkapazität f *(Batterie)*
   rated charging current   Nennladestrom m
   rated current   Nennstrom m
   rated frequency   Nennfrequenz f
   rated output   Nennleistung f
   rated speed   Nenndrehzahl f
   rated torque   Nenndrehmoment n
   rated voltage   Nennspannung f
ratio   Verhältnis n
   gear ratio   Übersetzungsverhältnis n
   power-to-mass ratio   Leistungsmasse f, leistungsbezogene Masse f
   (Triebfahrzeug)
   reverse ratio   Rückwärtsgang m

reaction rod Reaktionsstange f *(Radaufhängung)*
reaction time Ansprechzeit f, Reaktionszeit f
reactor core Reaktor-Brennkammer f *(Abgasreinigung)*
readiness indicator Bereitschaftsanzeige f
reading lamp Leseleuchte f
ready bereit, fertig
  ready for driving fahrbereit
  ready for operation betriebsbereit
  ready for service fahrbereit
  ready for use betriebsbereit
ream/to aufbohren *(Motor)*
rear rückwärtig. Hinter-, Rück-
  rear axle Hinterachse f
  rear axle housing Hinterachsengehäuse n
  rear axle tube Hinterachsrohr n
  rear compartment Fondraum m
  rear deck lid Kofferraumdeckel m, Kofferraumklappe f
  rear door Hecktür f, Rückwandtür f
  rear drive Heckantrieb m
  rear end Fahrzeugheck n, Heck n
  rear engine Heckmotor m
  rear flap Heckklappe f
  rear flasher lamp Schlußblinkleuchte f, Heckblinkleuchte f
  rear fog lamp Nebelschlußleuchte f
  rear lamp Schlußleuchte f, Rückleuchte f, Heckleuchte f
  rear licence plate lamp Schluß-Kennzeichenleuchte f
  rear light Rücklicht n, Schlußlicht n
  rear mirror Rückspiegel m, Außenspiegel m
  rear muffler Nachschalldämpfer m
  rear number-plate lamp hintere Kennzeichenleuchte f
  rear panel Rückwand f
  rear parcel shelf Hutablage f
  rear reflector Rückstrahler m
  rear seat Rücksitz m
  rear silencer Nachschalldämpfer m
  rear spoiler Heckspoiler m
  rear spring Hinterfeder f
  rear suspension Hinterachsaufhängung f, Hinterradaufhängung f
  rear view mirror Rückspiegel m

rear wall Rückwand f
rear wheel Hinterrad n
rear wheel drive Hinterradantrieb m, Heckantrieb m
rear wheel suspension Hinterradaufhängung f
rear window Heckscheibe f
rear window shelf Hutablage f
rear window wiper Heckscheibenwischer m
rebound travel Ausfederung f *(Stoßdämpfer)*
recess Aussparung f, Mulde f
recessed headlamp Einbauscheinwerfer m
reciprocating hin- und herbewegend
reciprocating piston engine Hubkolbenmotor m
recirculating ball steering Kugelumlauflenkung f
reclining seat verstellbarer Sitz m, Liegesitz m
recovery service Abschleppdienst m, Bergungsdienst m
recreational trailer Wohnanhänger m
recreational vehicle Wohnwagen m
rectangular headlamp Rechteckscheinwerfer m, rechteckförmiger
  Scheinwerfer m
rectifier Gleichrichter m
reducing valve Drosselventil n
reduction Verminderung f, Verringerung f, Untersetzung f
  reduction connection Reduzierstutzen m
  reduction ratio Untersetzungsverhältnis n
  reduction gear Vorgelege n *(Getriebe)*
  reduction valve Reduzierventil n
  final-reduction gear Radvorgelege n
  multi-step reduction gear Stufengetriebe n
refilling quantity Nachfüllmenge f
reflecting-type licence plate reflektierendes Kennzeichenschild n
reflect/to rückstrahlen
reflection Reflexion f, Spiegelung f
reflector Reflektor m, Rückstrahler m
  rear reflector Rückstrahler m
  reflex reflector Rückstrahler m, Katzenauge n
refrigerant Kühlmittel n, Kältemittel n
  refrigerant compressor Kühlmittelkompressor m,
  Kältekompressor m
refrigerated truck Kühlwagen m

refrigerator Kühlmaschine f, Kühlaggregat n
  refrigerator van Kühlwagen m
  refrigerator vehicle Kühlfahrzeug n
refuel/to tanken, auftanken
registration Zulassung f
  registration book Zulassungsschein m
  registration number Zulassungsnummer f, amtliches
  Kennzeichen n
  registration plate Kennzeichenschild n, Nummernschild n
regular normal
  regular gasoline Normalbenzin n
  regular-grade fuel (gasoline, petrol) Normalbenzin n
regulate/to regulieren, regeln, verstellen, lenken
regulating Verstell-, Einstell-, Regulier-
  regulating range Verstellbereich m
  regulating resistance Regulierwiderstand m
  regulating screw Regulierschraube f
  regulating spring Regulierfeder f, Einstellfeder f
  regulating valve Regulierventil n
regulator Regler m
  regulator for headlamp adjustment Leuchtweitenregler m
  pressure regulator Druckregler m
  voltage regulator Spannungsregler m *(Generator)*
  windshield regulator Fensteraussteller m
reinforced belt structure Gürtelaufbau m *(Reifen)*
reinforcement Verstärkung f
  reinforcement plate Verstärkungblech n
relay Relais n
  relay box Relaiskasten n
  relay valve Relaisventil n
  flasher relay Blinkrelais n
release/to auslösen, freigeben
  release the accelerator/to Gas wegnehmen
  release the clutch pedal/to einkuppeln
release Freigabe f, Auslösung f
  release cam Auslösenocken m
  release contact Auslösekontakt m
  release fork Ausrückgabel f, Kupplungsausrückgabel f
  release lever Rückzughebel m, Auslösehebel m, Ausrückhebel m,

Kupplungsausrückhebel m, Verstellhebel m
release shaft Ausrückwelle f, Kupplungsausrückwelle f
release sleeve Ausrückmuffe f, Kupplungsausrückmuffe f
release spring Auslösefeder f, Rückholfeder f, Rückzugfeder f,
Bremsbackenrückholfeder f, Bremsbackenrückzugfeder f
automatic release Selbstauslösung f
releasing lever Ausklinkhebel m
releasing time Abfallzeit f *(Relais)*
reliability Zuverlässigkeit f
reliability of operation Betriebszuverlässigkeit f
reliable betriebssicher; zuverlässig
relief valve Entlastungsventil n, Überdruckventil n
remote control Fernsteuerung f, Fernschaltung f
remo[u]ld/to runderneuern *(Reifen)*
repair Reparatur f
repair handbook Reparaturanleitung f
repair manual Werkstatthandbuch n
repair set Reparatursatz m
repair shop Reparaturwerkstatt f, Werkstatt f
repeating relay Wiederholrelais n
replacement Austausch m, Ersatz m
replacement engine Austauschmotor m
replacement part Ersatzteil n
reserve Reserve f
reserve fuel tank Kraftstoff-Reservebehälter n,
Reserve-Kraftstoffbehälter m
reserve tank Reservebehälter m
reservoir Vorratsbehälter m, Behälter m
auxiliary air reservoir Zusatzluftbehälter m
expansion reservoir Expansionsgefäß n
main air reservoir Hauptluftbehälter m *(Bremse)*
reset spring Rückstellfeder f
resistance Widerstand m, Beständigkeit f
abrasive resistance Abriebfestigkeit f
heat resistance Wärmebeständigkeit f
rolling resistance Laufwiderstand m, Rollwiderstand m,
Fahrwiderstand m
starting resistance Anlaßwiderstand m
resistant beständig

resistant to aging alterungsbeständig
acid-resistant säurebeständig
resistor Widerstand m *(Bauteil)*
  resistor-type spark plug Zündkerze f mit eingebautem Widerstand, Kerze f mit eingebautem Widerstand
  heater plug resistor Glühkerzenwiderstand m
resonance Resonanz f
resonator Nebenschalldämpfer m
  resonator intake silencer Reflexionsansauggeräuschdämpfer m
response Ansprechen n, Antwort f, Reaktion f
  response time Ansprechzeit f
rest Träger m, Lehne f, Stütze f
  rest position Ruhelage f
  foot rest Fußstütze f
restraint system Haltesystem n, Rückhaltesystem n, Haltevorrichtung f, Halteeinrichtung f
restricted air-bleed jet Luftkorrekturdüse f *(am Vergaser)*
restriction Verengung f
retainer Zierring m, Klemme f, Klinke f
  oil retainer Ölschutzblech n, Ölspritzblech n
  oil retainer ring Ölspritzring n
  ring retainer Kolbenringsicherung f, Ringsicherung f
retard/to verlangsamen, verzögern
retarded ignition Spätzündung f, Nachzündung f
retractor Aufroller m, Aufrolleinrichtung f, Aufrollautomatik f, Bandaufroller m
  retractor belt Automatikgurt m
  retractor force Aufrollkraft f, Rückzugskraft f
  retractor spring Rückzugfeder f, Rückholfeder f, Bremsbackenrückholfeder f, Bremsbackenrückzugfeder f
retread/to runderneuern *(Reifen)*
retroaction Rückwirkung f
return Rücklauf m, Rückführung f
  return line (pipe) Rücklaufleitung f
  return spring Rückholfeder f
  non-return valve Rückschlagventil n
  water return circuit Wasserrückleitung f
reverse rückwärts
  reverse gear Rückwärtsgetriebe n

reverse gear shaft Vorgelegewelle f
reverse idler gear Rücklaufrad n
reverse idler shaft Rücklaufachse f
reverse ratio Rückwärtsgang m
reverse scavenging Umkehrspülung f
reversing Umkehr-, Umlenk-
reversing gearbox Umkehrgetriebe n
reversing lamp Rückfahrlampe f, Rückfahrscheinwerfer m
reversing lever Umkehrhebel m, Umlenkhebel m
reversing projector Rückscheinwerfer m
reversing pulley Umlenkrolle f
revision Durchsicht f, Überarbeitung f
revolution Umdrehung f
revolution counter Tourenzähler m, Drehzahlmesser m
revolutions per minute Umdrehungen fpl pro Minute *(Drehzahl)*
axis of revolution Rotationsachse f
number of revolutions Umdrehungszahl f
rheostat Regelwiderstand m
rib Rippe f, Lamelle f *(Kühler)*; Stollen m *(Reifen)*
reinforcing rib Verstärkungsrippe f
ribbed radiator Lamellenkühler m
right rechts; Recht n
right-hand drive Rechtslenkung f
right-hand steering Rechtslenkung f
right-hand turn Rechtsabbiegen n
right-hand traffic Rechtsverkehr m
right of way Vorfahrtsrecht n
turn to the right/to nach rechts abbiegen
rigid starr
rigid axle Starrachse f
rigid beam axle Starrachse f
rim Felge f, Radkranz m
rim bead Felgenrand m
rim diameter Felgendurchmesser m
rim drop center Felgentiefbett n
rim flange Felgenhorn n, Hornring m
rim ring Felgenring m
rim tape Felgenband n
rim taper Wulstsitzwinkel m

rim tool Felgenabziehhebel m, Reifenheber m,
Reifenmontierhebel m
rim well Felgenbett n
base rim Grundfelge f, Felgenbett n, Grundpartie f, Basisteil n
clincher rim Wulstreifen m
drop-center rim Tiefbettfelge f
flat-base rim Flachbettfelge f
flywheel rim Schwungradkranz m, Schwungkranz m
toothed rim Zahnkranz m
toothed wheel rim Zahnkranz m
well-base rim Tiefbettfelge f
ring Ring m, Öse f, Kolbenring m
  ring carrier Ringträger m
  ring clip Ringfeder f
  ring clogging Verkleben n der Kolbenringe
  ring gap Stoßöffnung f, Stoß m
  ring gear Zahnkranz m, Tellerrad n, Hohlrad n
  ring groove Ringnut f, Kolbenringnut, Kolbennut f
  ring lock Ringsicherung f
  ring spanner Ringschlüssel m
  ring sticking Festkleben n (Hängenbleiben) der Kolbenringe,
  Ringstecken n, Ringverkleben n
  adjusting ring Stellring m
  axial bearing ring Axiallagerring m
  clamping ring Klemmring m
  cover ring Abdeckring m
  crown ring Starterzahnkranz m
  cutting ring Schneidring m
  gear ring Zahnkranz m
  inner ring Innenring m
  insulating ring Isolierring m
  locking ring Sicherungsring m, Sperring m
  loop-ring Rundschnurring m
  mounting ring Montagering m
  oil retainer ring Ölspritzring m
  oil ring Schmierring m
  oil scraper ring Ölabstreifring m
  piston ring Kolbenring m
  rim ring Felgenring m

scraper ring Abstreifring m
sheet metal ring Blechring m
snap ring Sprengring m
snow chain ring Schneekettenring m
spacer ring Distanzring m
starter ring gear Anlasserzahnkranz m
stop ring Anschlagring m, Arretierring m
synchro ring Synchronring m
thrust ring Anpreßring m
toothed flywheel ring Zahnkranz m
trailer ring Anhängeröse f
yoke ring Jochring m
rinse/to spülen
rivet Niet m
rivet tool Nietwerkzeug n
flat rivet Flachniet m
flat-round rivet Flachrundniet m
hollow rivet Hohlniet m
riveting Niet-
riveting device Nietvorrichtung f
riveting pin Nietstift m
road Straße f, Landstraße f
road clearance Bodenfreiheit f, Bodenabstand m
road drag Rollwiderstand m
road friction Fahrbahnreibung f
road grip Kraftschluß m zur Straße *(Laufstreifen beim Reifen)*
road holding Straßenlage f
road-holding straßengängig
road holding quality Straßenlage f
road resistance Rollwiderstand m
road safety Verkehrssicherheit f
road test Straßenversuch m, Fahrversuch m
road wheel Laufrad n, Rad n
off-the-road geländegängig, Gelände-
roadster offener Zweisitzer m
roadworthy verkehrstüchtig
rocker Kipphebel m, Schlepphebel m
rocker arm Schwinghebel m, Kipphebel m, Schlepphebel m
rocker arm shaft Kipphebelachse f, Kipphebelwelle f,

Schwinghebelachse f
rocker arm top link   Kipphebelquerlenker m
rocker clearance   Ventilspiel n
rocker cover   Zylinderkopfhaube f, Ventildeckel m,
Ventilabdeckung f
rocker lever   Kipphebel m, Schlepphebel m, Schwinghebel m
rocker panel   Einstiegsblech n, Trittbrett n
rocker shaft   Kipphebelachse f, Kipphebelwelle f,
Schwinghebelachse f
rocker switch   Kippschalter m
rod   Stange f, Strebe f
  adjusting rod   Einstellstange f
  connecting rod   Pleuel n, Pleuelstange f, Verbindungsstange f
  connecting rod bearing   Pleuellager n
  connecting rod bush   Pleuelbüchse f
  control rod   Betätigungsstange f
  coupling rod   Kupplungsstange f, Kuppelstange f
  cross steering rod   Querlenkstange f
  disengaging rod   Ausrückstange f
  drag rod   Lenkschubstange f, Spurstange f
  driving rod   Antriebsstange f
  guide rod   Führungsstange f
  link rod   Gelenkstange f
  Panhard rod   Panhard-Stab m, Querlenker m
  piston rod   Kolbenstange f
  pump push rod   Pumpenstößel m
  push rod   Schubstange f, Stoßstange f, Stößelstange f
  push rod guide   Stoßstangenführung f
  radius rod   Lenkzwischenstange f, Achsstrebe f, Zugstange f
  steering rod   Lenkstange f, Spurstange f
  steering rod lever   Spurstangenhebel m
  tie rod lever   Zugstangenhebel m
  torsion rod   Torsionsstab m
  towing rod   Abschleppstange f
  track rod   Spurstange f
  transverse rod   Querlenker m
  valve push rod   Ventilstoßstange f, Ventilstößelstange f
roll   Wanken n, Rollen n, Schlingern n
  roll bar   Überrollbügel m; Stabilisator m, Querstabilisator m,

Drehstab m
roll stabilizer Stabilisator m, Querstabilisator m, Drehstab m
roller Rolle f, Walze f
  roller cell pump Rollenzellenpumpe f
  roller bearing Rollenlager n, Wälzlager n
  roller blind Rolljalousie f, Rollo n, aufrollbare Jalousie f,
  aufrollbare Sonnenblende f
  roller cage Rollenkäfig m
  roller chain Rollenkette f
  roller dynamometer Rollenprüfstand m
  roller tappet Rollenstößel m
  axial roller bearing Axialrollenlager n
  bevel roller bearing Kegelrollenlager n
  chain tension roller Kettenspannrolle f
  cylinder roller bearing Zylinderrollenlager n
  guide roller Führungsrolle f, Lenkrolle f, Leitrolle f
  pressure roller Druckwalze f, Druckrolle f, Anpreßrolle f
  self-aligning roller bearing Pendelrollenlager n
rolling Roll-
  rolling circumference Abrollumfang m
  rolling friction Rollreibung f, Rollwiderstand m
  rolling moment Wankmoment n, Rollmoment n
  rolling radius Rollradius m, Rollhalbmesser m, Halbmesser m des
  belasteten Reifens
  rolling resistance Laufwiderstand m, Rollwiderstand m,
  Fahrwiderstand m
rollover/to überrollen, überschlagen, umkippen
rollover bar Überrollbügel m
roof Dach n
  roof bow Dachspriegel m
  roof drip mo[u]lding Dachrinne f, Dachleiste f
  roof fold Dachfalz m
  roof frame Dachrahmen m
  roof lamp Deckenleuchte f, Innenlampe f
  roof lining Dachauskleidung f, Himmel m, Dachverkleidung f
  roof luggage carrier Dachgepäckträger m
  roof panel Dachbeplankung f, Dachhaut f, Dachblech n
  roof panel[l]ing Dachbeplankung f, Dachblech n
  roof pillar Dachsäule f, Dachpfosten m

roof post   Dachpfosten m, Dachsäule f
roof rack   Gepäckrost m, Dachgepäckträger m
roof rail   Regenrinne f
roof skin   Dachbeplankung f, Dachhaut f, Dachblech n
roof stiffness   Dachsteifigkeit f
roof top carrier   Dachgepäckträger m
inside roof lining   Himmel m, innere Deckenauskleidung f
sliding roof   Schiebedach n
roots blower   Rootsgebläse n
rope   Seil n
  rope drum   Seiltrommel f
  rope guide   Seilführung f
  rope pulley   Seilrolle f
rotary   Dreh-, Kreis-, Rotations-
  rotary combustion engine   Kreiskolbenmotor m,
Rotationskolbenmotor m, Drehkolbenmotor m, Wankel-Motor m
  rotary current   Drehstrom m
  rotary device   Drehvorrichtung f
  rotary disk valve   Drehschieber m
  rotary engine   Kreiskolbenmotor m, Drehkolbenmotor m,
Rotationskolbenmotor m, Wankel-Motor m
  rotary knob   Drehknopf m
  rotary movement   Umlaufbewegung f
  rotary piston engine   Kreiskolbenmotor/KKM m,
Drehkolbenmotor m, Rotationskolbenmotor m, Wankel-Motor m
  rotary pump   Kreiselpumpe f
  rotary slide   Drehschieber m
  rotary slide valve   Drehschieber m
  rotary switch   Drehschalter m
  rotary wiper motor   Umlaufwischermotor m
rotating flashing beacon   Rundumkennleuchte f
rotation   Umdrehung f, Drehung f
  anti-clockwise rotation   Linksdrehung f, Linkslauf m
  clockwise rotation   Rechtsdrehung f
  left-hand rotation   Linkslauf m, Linksdrehung f
  right-head rotation   Rechtslauf m
  sense of rotation   Drehrichtung f
rotational   Rotations-, Umlauf-, Dreh-
  rotational axis   Rotationsachse f

rotational speed Umlaufgeschwindigkeit f, Drehgeschwindigkeit f
rotor Rotor m, Läufer m, Flügelrad m *(Kreiskolbenmotor)*;
Kolben m, Kreiskolben m
rotor arm Verteilerfinger m, Verteilerläufer m,
Zündverteilerfinger m
rotor bank Kolbeneinheit f, Rotoreinheit f
rotor contour (face, flank) Kolbenflanke f, Rotorflanke f,
Flanke f
rotor housing Mantel m, Kolbengehäuse n, Rotorgehäuse n,
Trochoidengehäuse n
rotor housing seal Manteldichtung f, Kolbengehäusedichtung f,
Rotorgehäusedichtung f
rotor recess Kolbenmulde f, Kolbenfenster n, Rotormulde f,
Rotorfenster n
rotor shaft Rotorwelle f
rotor tip seal Dichtleiste f, Scheitelleiste f, Radialdichtleiste f
distributor rotor Verteilerläufer m, Verteilerfinger m
distributor rotor arm Verteilerfinger m
magnetic rotor Magnetläufer m
short-circuit rotor Kurzschlußläufer m
twin-rotor Wankel engine Doppelläufer-Wankelmotor m
round rund
round plug socket Rundsteckhülse f
flat-round rivet Flachrundniet m
half-round rivet Halbrundniet m
rubber Gummi m, Profilgummi m
rubber bearing Gummilager n
rubber bellows Gummibalg m
rubber buffer Gummipuffer m
rubber bush Gummibüchse f, Gummitülle f
rubber covering Gummibelag m, Gummiverkleidung f
rubber damper Gummidämpfer m
rubber doughnut Gelenkscheibe f, Scheibengelenk n
rubber gasket Gummidichtung f
rubber grommet Gummitülle f
rubber hose Gummischlauch m
rubber insert Gummieinlage f
rubber joint Gummidichtung f
rubber lip Gummilippe f

rubber mat   Gummimatte f
rubber metal   Schwingmetall n
rubber mounting   Gummilager n
rubber pad   Gummiauflage f
rubber plug   Gummistopfen m
rubber sealing lip   Gummilippe f
rubber sleeve   Gummihülse f, Gummimanschette f
rubber socket   Gummimuffe f, Gummitülle f
rubber stop   Anschlaggummi m, Gummianschlag m
rubber strip   Gummileiste f
rubber valve   Gummiventil n
damping rubber   Dämpfungsgummi m, Stoßdämpfergummi m
foam rubber   Schaumgummi m
foam rubber profile   Moosgummiprofil n
profile rubber   Formgummi m
sealing rubber   Abdichtgummi m
solid-rubber tire   Vollgummireifen m
wiper rubber   Wischergummi m
run/to   fahren
  run-on/to   nachlaufen, nachdieseln
  run on petrol/to   mit Benzin fahren
run   Lauf m
  run-down battery   entladene Batterie f
  endurance run   Dauerlauf m
runner   Laufrolle f
running   Lauf-, Fahr-
  running board   Trittbrett n
  running gear   Fahrgestell n, Chassis n, Fahrwerk n
  running-in oil   Einlauföl n
  running-in period   Einlaufzeit f
  running plate   Trittplatte f
rust/to   rosten
rust   Rost m
  rust formation   Rostbildung f
  rust inhibitor (preventive)   Rostschutzmittel n,
  Rostschutz-Additiv n
  rust-protective paint   Rostschutzfarbe f
  rust remover   Entroster m, Rostentferner m,
  Rostentfernungsmittel n, Entrostungsmittel n

rustproof rostbeständig
  rustproof oil Korrosionsschutzöl n
saddle Sattel m
safety Sicherheit f
  safety belt Sicherheitsgurt m, Sitzgurt m
  safety bolt Sicherungsbolzen m
  safety cage Sicherheitszelle f *(Fahrgastraum)*
  safety catch lever Sicherungshebel m
  safety cell Sicherheitszelle f *(Fahrgastraum)*
  safety glass Sicherheitsglas n, Einschicht-Sicherheitsglas n
  safety lock Sicherungsschloß n
  safety nut Sicherungsmutter f
  safety roll bar Überrollbügel m
  safety sheet Sicherungsblech n
  safety valve Sicherheitsventil n
  road safety Verkehrssicherheit f
saloon Limousine f
salt spray test Salzsprühtest m
sample Probe f, Muster n
sash belt Oberkörpergurt m, Schultergurt m
saturation Sättigung f
scale Skala f, Maßstab m
  adjusting scale Einstellskala f
scavenge/to spülen
  scavenge port Spülschlitz m *(Schieber)*
scavenging Spülung f, Spül-
  scavenging air Spülluft f
  scavenging loop Umkehrspülung f
  scavenging pressure Spüldruck m
  scavenging pump Spülpumpe f
  loop scavenging Umkehrspülung f
  reverse scavenging Umkehrspülung f
scissors-type jack Scherenwagenheber m
scooter Roller m, Motorroller m
score mark Riefe f
scoring Riefenbildung f
scrap Schrott m
  scrap yard Schrottplatz m, Autofriedhof m
scrape/to schaben, abkratzen

scraper   Abstreifer m
  scraper ring   Abstreifring m
  oil scraper ring   Ölabstreifring m
screen   Abschirmhaube f, Sieb n, Windschutzscheibe f
  clear vision screen   Klarsichtscheibe f
  fresh air screen   Frischluftgitter n
  oil screen   Ölsieb n
screened cable   abgeschirmtes Kabel n
screening   Abschirmung f, Siebung f
  screening cap   Entstörkappe f
  screening plate   Abschirmblech n
  screening socket   Entstörstecker m
screenjet   Scheibenspüler m, Scheibenwascher m
screenwasher   Scheibenwaschanlage f
screw/to   schrauben
  screw off/to   abschrauben
  screw on/to   anschrauben
screw   Schraube f
  screw-and-nut steering gear   Lenkgetriebe n mit
  Schraubenspindel und Mutter, Schraubenlenkung f,
  Spindellenkgetriebe n
  screw bolt   Schraubenbolzen m
  screw cap   Gewindestopfen m, Schraubverschluß m,
  Verschlußschraube f
  screw connection   Schraubanschluß m, Schraubverbindung f
  screw driver   Schraubenzieher m
  screw-in socket   Einschraubstutzen m
  screw lifting jack   Wagenheber m, Schraubenheber m
  screw neck   Schraubstutzen m
  screw plug   Verschlußstopfen m, Verschlußschraube f
  screw-push starter   Schubschraubtriebanlasser m
  adjusting screw   Einstellschraube f
  air-vent screw   Entlüftungsschraube f *(Batterie)*
  cable terminal screw   Kabelklemmschraube f
  cap screw   Überwurfschraube f
  clamping screw   Klemmschraube f
  collar screw   Bundschraube f
  connecting screw   Anschlußschraube f
  countersunk screw   Senkschraube f

cylinder head screw   Zylinderkopfschraube f
fastening screw   Befestigungsschraube f
head screw   Kopfschraube f
hexagon screw   Sechskantschraube f
hexagon socket screw   Innensechskantschraube f, Imbusschraube f
hollow screw   Hohlschraube f
knurled head screw   Rändelschraube f
locking screw   Verschlußschraube f
ring screw   Ringschraube f
self-tapping screw   selbstschneidende Schraube f
setting screw   Einstellschraube f
sheet metal screw   Blechschraube f
stop screw   Anschlagschraube f
welding screw   Schweißschraube f
wheel screw   Radschraube f
screwing   Verschraubung f
scuff/to   scheuern, reiben
scuttle   Windlaufquerteil n, Querstrebe f *(über der Frontscheibe)*,
Stirnwand f, Spritzwand f
seal   Dichtung f, Abdichtung f
seal paste   Dichtungsmasse f
sealed beam headlamp   Sealed-beam-Scheinwerfer m *(eine
amerikanische Ausführung, bei der Lichtaustrittscheibe, Reflektor und
Glühlampe eine Einheit bilden)*
sealing   Dichtung f
sealing compound   Dichtmittel n
sealing lip   Dichtlippe f
sealing material   Dichtungsmaterial n
sealing rubber   Abdichtgummi m
sealing washer   Dichtring m
seam   Naht f, Schweißnaht f
searchlight   Suchscheinwerfer m
seat   Sitz m
seat adjuster   Sitzversteller m
seat adjustment   Sitzverstellung f
seat back   Rückenlehne f, Sitzlehne f, Sitzrückenlehne f
seat back adjustment   Sitzlehnenverstellung f,
Rücklehnenverstellung f
seat backrest   Sitzrückenlehne f, Sitzlehne f, Rückenlehne f

seat belt   Sitzgurt m, Sicherheitsgurt m

seat belt anchorage point   Sitzgurtverankerungspunkt m, Gurtverankerungspunkt m

seat belt buckle   Gurtschloß n, Gurtverschluß m

seat belt retractor   Aufroller m, Aufrolleinrichtung f, Aufrollvorrichtung f, Aufrollautomatik f, Bandaufroller m

seat belt system   Sitzgurtsystem n

seat bench   Sitzbank f

seat bucket   Sitzschale f

seat cover   Sitzbezug m, Sitzschonbezug m, Schonbezug m

seat cushion   Sitzpolster n, Sitzkissen n

seat cushion spring   Federeinlage f

seat fabric   Sitzbezug m

seat frame   Sitzgestell n

seat guide rail   Sitzlaufschiene f, Sitzschiene f

seat pan   Sitzschale f

seat rail   Sitzlaufschiene f, Sitzschiene f

seat springing   Sitzfederung f

seat upholstery   Sitzpolsterung f, Sitzpolster n, Polsterung f

driver's seat   Fahrersitz m

front seat   Vordersitz m

occasional seat   Notsitz m

reclining seat   verstellbarer Sitz m, Liegesitz m

spring seat   Federsitz m, Federteller m

valve seat   Ventilsitz m

secondary   Sekundär-, Neben-, Zweit-

secondary air   Sekundärluft f, Nebenluft f, Zusatzluft f

secondary air injection   Sekundärlufteinblasung f

secondary barrel   zweite Stufe f

secondary coil   Sekundärwicklung f

secondary combustion chamber   Nebenkammer f *(Motor)*

secondary shaft   Sekundärwelle f, Vorgelegewelle f

secondary stage   zweite Stufe f

secondary throat   Hauptlufttrichter m, Hauptzerstäuber m *(am Vergaser)*

secondary venturi   Nebenlufttrichter m, Mischkammer f

secondary winding   Sekundärwicklung f

section   Schnitt m

wire cross section   Leitungsquerschnitt m

secure/to befestigen, sichern
sedan Limousine f
sediment Ablagerung f
  sediment bowl Schmutzabscheider m
segment Lamelle f, Segment n
  collector segment Kollektorlamelle f
  coupling segment Kupplungslamelle f
seize/to festfressen, verklemmen *(z.B. Kolben)*
selector Gangwähler m, Schaltgriff m
  selector cable Schaltkabel n
  selector control Wählhebel m, Getriebewählhebel m
  selector fork Schaltgabel f
  selector lever Wählhebel m, Getriebewählhebel m
  selector lever position indicator
  Wählhebelstellungsanzeiger m
  selector rod Schaltstange f
  gear selector mechanism Gangwähler m
  slide selector shaft Schaltstange f
self Selbst-
  self-acting selbstständig, selbsttätig, automatisch
  self-adjusting selbstnachstellend, selbsteinstellend
  self-aligning ball bearing Pendelkugellager n
  self-aligning bearing Pendellager n
  self-aligning roller bearing Pendelrollenlager n
  self-aligning torque Rückstellmoment n *(Lenkung)*
  self-cancel[l]ing selbstrückstellend, selbstausschaltend
  self-ignition Eigenzündung f, Selbstzündung f
  self-induction Selbstinduktion f
  self-locking selbstsperrend, selbsthemmend
  self-locking differential selbstsperrendes Ausgleichgetriebe n,
  selbstsperrendes Differential n
  self-sealing selbstdichtend, selbstabdichtend
  self-supporting body selbsttragende Karosserie f
  self-tapping screw Gewindeschneidschraube f
semi Halb-
  semi-automatic transmission halbautomatisches Getriebe n
  semi-charged halbgeladen
  semiconductor Halbleiter m
  semi-downdraft carburetor Schrägstromvergaser m

semitrailer Sattelanhänger m, Sattelauflieger m, Auflieger m
semitrailer train Sattelschlepperzug m, Sattelzug m,
Sattelschlepper m
semitrailer unit Sattelschlepperzug m, Sattelzug m,
Sattelschlepper m
semi-trailing arm Schräglenker m
semi-trailing arm axle Schräglenkerachse f, Diagonalpendelachse f
sense of rotation Drehrichtung f
sensing time Ansprechzeit f
sensor Sensor m, Fühler m
separate getrennt, Fremd-
separate air connection Fremdluftanschluß m
separate gear change getrennte Schaltung f
separator Separator m, Abscheider m; Trennplatte f *(Batterie)*
oil separator Ölabscheider m
water separator Wasserabscheider m
series Serie f, Baureihe f; Reihe f
series connection Reihenschaltung f, Reihenschluß m
series production Serienfertigung f, Serienproduktion f
serrated zackenförmig, kerbverzahnt
serration Kerbverzahnung f
service Betrieb m, Kundendienst m
service brake Betriebsbremse f, Betriebsbremsanlage f,
Fußbremse f
service brake system Betriebsbremsanlage f, Betriebsbremse f
service interval Kundendienstintervall n
service life Lebensdauer f, Laufzeit f
service manual Kundendienstheft n
servo brake Servobremse f, Bremse f mit Bremskraftverstärker
set 1. Aggregat n, Anlage f, Satz m; 2. Stell-, Einstell-
set screw Stellschraube f
repair set Reparatursatz m
tool set Werkzeugsatz m
setting Einstellung f
setting angle Einstellwinkel m
setting mark Einstellmarke f
setting screw Einstellschraube f
ignition setting Zündeinstellung f
spark setting Zündeinstellung f

shackle   Lasche f, Schäkel m
shaft   Welle f
  shaft bearing   Wellenlager n
  shaft bush   Wellenbuchse f
  shaft pivot   Wellenzapfen m
  shaft ring   Wellendichtring m
  shaft sealing ring   Wellendichtring m
  shaft speed   Wellendrehzahl f
  shaft with pinion   Welle f mit Ritzel
  cam shaft   Nockenwelle f
  cardan shaft   Kardanwelle f
  centering shaft   Zentrierwelle f
  crank shaft   Kurbelwelle f
  differential pinion shaft   Ausgleichradachse f, Differentialachse f
  differential shaft   Differentialwelle f
  disengaging shaft   Ausrückwelle f
  distributor shaft   Verteilerwelle f
  drive (driving) shaft   Antriebswelle f, Kardanwelle f, Gelenkwelle f
  gear shaft   Zahnradwelle f
  half shaft   Halbachse f, Achswelle f
  idler shaft   Vorgelegewelle f
  intermediate shaft   Zwischenwelle f
  pinion drive shaft   Ritzelwelle f
  propulsion shaft   Antriebswelle f
  rear wheel shaft   Hinterradantriebswelle f
  reverse idler shaft   Rücklaufachse f, Rücklaufwelle f
  rocker lever shaft   Kipphebelwelle f
  secondary shaft   Sekundärwelle f, Vorgelegewelle f
  slide selector shaft   Schaltstange f
  supporting shaft   Tragwelle f
  toothed shaft   Zahnwelle f
  transmission shaft   Antriebswelle f
shear   Scher-, Abscher-
  shear force   Scherkraft f
  shear pin   Scherstift m, Scherbolzen m
  shear strength   Abscherfestigkeit f
sheet   Blech n, Tafel f
  sheet metal   Blech n
  sheet metal radiator   Lamellenkühler m

sheet metal screw   Blechschraube f
sheet steel   Stahlblech n
air guide sheet   Luftleitblech n
covering sheet   Abdeckblech n
dust sheet   Staubblech n
front wall sheet   Vorderwandblech n
gusset sheet   Eckblech n, Knotenblech n
oil deflector sheet   Ölabstreifblech n
spacer sheet   Distanzblech n
supporting sheet   Stützblech n, Tragblech n
sheeting   Blechverkleidung f
shelf   Ablageschale f, Ablagefach n
shell   Gehäuse n, Schale f
radiator shell   Kühlergehäuse n
spark plug shell   Zündkerzengehäuse n
shield   Schild m, Schutzschild m
frost shield   Heizscheibe f
spark-plug shield   Funkenabschirmung f
shielding   Abschirmung f; Entstörung f *(z.B. Zündung)*
radio-shielding   Entstörung f, Störschutz m
radio-shielding cable   Enstörkabel n
radio-shielding capacitor   Entstörkondensator m
radio-shielding plug   Entstörstecker m
shift/to   schalten *(Gang)*
shift down/to   herunterschalten, zurückschalten *(Gang)*
shift up/to   heraufschalten *(Gang)*
shift   Schaltvorgang m, Gangwechsel m *(s.a. shifting)*
shift axle   Schaltachse f
shift cable   Schaltkabel n
shift control   Gangschaltung f, Schalthebel m, Schaltknüppel m, Gangschalthebel m
shift control housing   Schaltgehäuse n
shift dog   Schaltklaue f
shift fork   Schaltgabel f
shift indicator   Ganganzeige f, Wählhebelstellungsanzeige f
shift knob   Schalthebelknopf m
shift lever   Schalthebel m, Schaltknüppel m, Gangschalthebel m
shift lever bracket   Schaltbock m
shift lever knob   Schalthebelknopf m

shift linkage   Schaltgestänge n
shift **mechanism**   Schaltmechanismus m, Gangschaltung f
shift **pattern**   Schaltschema n, Ganganordnung f
shift point   Schaltpunkt m, Schaltzeitpunkt m
shift rail   Schaltschiene f, Schaltstange f
shift rod   Schaltstange f
shift sleeve   Schaltmuffe f
shift valve   Schaltventil n, Schaltschieber m
compressed-air shift cylinder   Druckluftschaltzylinder m
gear-shift bar   Schaltstange f
gear-shift control mechanism   Gangschaltung f
gear-shift cover   Schaltdeckel m
gear-shift fork   Schaltgabel f
gear-shift lever   Schalthebel m, Handschalthebel m
gear-shift-lever shaft   Schaltwelle f
gear-shift mechanism   Gangschaltung f
gear-shift pattern   Ganganordnung f
gear-shift rail   Schaltschiene f
gear-shift sleeve   Schaltmuffe f
gear shift starter   Schubtriebanlasser m
screw shift pinion starter   Schubschraubtriebanlasser m
shifter   Schalthebel m, Schaltknüppel m, Ganghebel m
shiftless transmission   automatisches Getriebe n
shifting   Schalt- *(s.a. shift)*
shifting arm   Schaltarm m
shifting bar   Schaltriegel m, Schaltstange f
shifting cable   Schaltkabel n
shifting fork   Schiebegabel f, Schaltgabel f
shifting gate   Schaltkulisse f, Kulisse f
shifting shaft   Schaltwelle f
shifting travel   Schaltweg m
ball and socket type gear shifting   Kugelschaltung f
shim   Unterlegscheibe f
shimmy   Radflattern n
wheel shimmy   Radflattern n
shock   Stoß m
shock absorber   Stoßdämpfer m
shock absorber piston   Stoßdämpferkolben m
shock absorber plunger   Stoßdämpferkolben m

shock absorber spring Stoßfeder f
shock absorber strut Stoßdämpferbein n, Dämpferbein n
shock absorber tube Stoßdämpferrohr n
telescopic shock absorber Teleskopstoßdämpfer m
shoe Backe f, Bremsbacke f
  shoe clutch Backenkupplung f
  shoe-type brake Backenbremse f
  cable shoe Kabelschuh m
  floating shoe Gleitbacken m *(Bremse)*
  inside shoe brake Innenbackenbremse f
  pivoted shoe Drehbacken m *(Bremse)*
  trailing shoe Aufbaubacken m *(Bremse)*, Ablaufbacke f
short kurz
  short circuit Kurzschluß m
  short-circuit brake Kurzschlußbremse f
  short-circuit line Kurzschlußleitung f
  short-circuit valve Kurzschlußventil n
  short-distance haulage Güternahverkehr m, Nahtransport m,
  Nahverkehr n, Kurzstreckenbetrieb m, Kurzstreckenverkehr m
  short-distance interference suppression Nahentstörung f
  short-distance radio-shielding Nahentstörung f
  short-distance traffic Kurzstreckenverkehr m,
  Kurzstreckenbetrieb m
  short-distance transportation Güternahverkehr m,
  Nahtransport m, Nahverkehr m
  short-stroke engine Kurzhubmotor m
shorted coil Windungsschluß m
shoulder Schulter f, Reifenschulter f
  shoulder bearing Schulterlager n
  shoulder harness Schultergurt m, Sicherheitsgurt m
  shoulder-lap belt system Dreipunktgurt m
shovel loader Schaufellader m
shunt Nebenschluß m
  shunt resistance Nebenschlußwiderstand m *(Eigenschaft)*
  shunt resistor Nebenschlußwiderstand m *(Bauelement)*
  shunt-wound motor Nebenschlußmotor m
shut-off valve Abschaltventil n, Sperrventil n
side Seite f
  side-car Seitenwagen m, Beiwagen m

side clearance Nutspiel n *(Kreiskolbenmotor)*; Axialspiel n,
Kolbenringspiel n, Kolbenringlängsspiel n
side cover Seitendeckel n
side gate Seitenwand f
side housing Seitenteil n
side inlet port Seiteneinlaß m
side lamp Begrenzungsleuchte f, Seitenleuchte f, Positionsleuchte f
side light Begrenzungslicht n
side marker Peilstange f
side-marker lamp Seiten*markierungs*leuchte f,
Begrenzungsleuchte f
side member Längsträger m, Längsholm m, Seitenträger m,
Rahmenlängsträger m
side mirror Außenspiegel m, Außenrückspiegel m
side panel Seitenwand f
side panel[l]ing Seitenverkleidung f, Seitenwandverkleidung f
side rack Seitenwand f
side rail Rahmenlängsträger m, Längsträger n, Längsholm m,
seitlicher Längsträger m
side ring Seitenring m
side seal Dichtstreifen m
side shaft Abtriebswelle f
side sill Türschweller m
side tipper Seitenkipper m
side-wind immunity Seitenwindunempfindlichkeit f
side-wind sensitivity Seitenwindempfindlichkeit f
side window Seitenfenster f, Seitenscheibe f
sidedraft carburetor Flachstrom-Vergaser m, Horizontal-Vergaser m
sidewall Seitenwand f
  sidewall rib Scheuerrippe f, Scheuerleiste f
sieve Sieb n
  filter sieve Filtersieb n
  radiator sieve Kühlersieb n
signal Signal n, Zeichen n
  signal button Signalknopf m
  signal horn Signalhorn n, Hupe f
  signal push button Signaldruckknopf m
  signal transmitter Signalgeber m
  flash signal Lichthupe f, Blinkzeichen n

silencer Abgasschalldämpfer m, Schalldämpfer m, Auspufftopf m
  exhaust silencer Auspufftopf m, Auspuffschalldämpfer m,
  Abgasschalldämpfer m
  intake silencer Ansauggeräuschdämpfer m
silentblock Silentblock m
sill Ladekante f, Schwelle f
  door sill Türschwelle f
single Einzel-, Einfach-
  single-arm Querlenker m ohne Längsführung
  single-axle suspension Einzelachsaufhängung f,
  Einzelachsfederung f, Einachsfederung f
  single-barrel carburetor Einfachvergaser m, Einzelvergaser m
  single-bore carburetor Einfachvergaser n, Einzelvergaser m
  single-chamber brake cylinder Einkammerbremszylinder m
  single-choke carburetor Einfachvergaser m, Einzelvergaser m
  single-circuit brake system Einkreisbremsanlage f
  single-coil ignition Einspulenzündung f
  single-core cable einadriges Kabel n
  single-disk clutch Einscheibenkupplung f
  single-jet Einfachdüse f
  single-piece rim einteilige Felge f
  single-plate clutch Einscheibenkupplung f
  single-pole einpolig
  single-speed blower (fan) einstufiges Gebläse n, einstufiger
  Ventilator m
  single-stage einstufig
  single-stage torque converter Einstufen-Drehmomentwandler m
  single-wheel drive Einzelradantrieb m
sintered metal Sintermetall n
sipe Lamelle f; Profilrille f
siped tire feinprofilierter Reifen m
sipes Feinprofilierung f, Lamellierung f
six sechs
  six-bearing sechsfach gelagert
  six wheeler Dreiachser m
skid/to rutschen, schleudern
skid Lauffläche f *(Reifen)*
  skid depth Profiltiefe f *(Reifen)*
  skid mark Bremsspur f

skid resistance   Rutschsicherheit f, Reibungswiderstand m
skirt   Schaft m, Kolbenschaft m, Kolbenmantel m
  skirt ring   Ölabstreifring m am Kolbenschaft
slack   Totgang m, freies Spiel n
  slack adjuster   Leerwegeinsteller m *(Bremse)*
slap   Schlag m *(Kolben)*
slave cylinder   Nebenzylinder m; Radzylinder m,
  Radbremszylinder m
sleeve   Büchse f, Manschette f, Hülse f, Muffe f
  sleeve valve   Schieber m
  sleeve valve engine   Schiebermotor m
  bearing sleeve   Lagerhülse f
  cable sleeve   Kabelmuffe f, Kabelhülse f
  clamping sleeve   Klemmuffe f, Klemmhülse f
  clutch release sleeve   Ausrückmuffe f
  coupling sleeve   Kupplungsmuffe f
  gear-shift sleeve   Schaltmuffe f
  insulating sleeve   Isolierbuchse f, Isolierhülse f
  joint sleeve   Abdichthülse f
  rubber sleeve   Gummihülse f, Gummimanschette f
  shift sleeve   Schaltmuffe f
  stop sleeve   Anschlaghülse f
slick   Slick m, profilloser Rennreifen m
slide   Schlitten m, Schieber m, Gleit-, Schub-
  slide bearing   Gleitlager n
  slide face   Gleitfläche f
  slide rail   Gleitschiene f, Laufschiene f
  slide selector shaft   Schaltstange f
  slide valve   Schieber m, Flachschieber m
  slide valve engine   Schiebermotor m
  slide valve gear   Schiebertrieb m, Schiebersteuerung f
  flat slide valve   Flachschieber m
slider   Schieber m
sliding   Gleit-, Schub-, Schieb-
  sliding armature starter   Schubankeranlasser m
  sliding bearing   Gleitlager n
  sliding block   Gleitstein m
  sliding contact   Schleifkontakt m, Gleitkontakt m
  sliding door   Schiebetür f

sliding gear   Schieberad n, verschiebbares Zahnrad n
sliding gear starter motor   Schubtrieb-Anlasser m
sliding jaw   Gleitbacke f
sliding joint   Schiebegelenk n, Gleitgelenk n
sliding rail   Gleitschiene f
sliding rod   Schieberstange f
sliding roof   Schiebedach n, Kurbeldach n
sliding seat   verstellbarer [einstellbarer] Sitz m, Schiebesitz m
sliding selector shaft   Schaltstange f
sliding shaft   Schiebewelle f
sliding sun-roof   Schiebedach n
sliding top   Schiebedach n
slinger ring   Schleuderring m
slip   Schlupf m, Radschlupf m, Reifenschlupf m
slip angle   Schräglaufwinkel m
slip ring   Schleifring m
slip stream   Fahrtwind m
slip universal joint   Gleitkreuzgelenk n
slip/to   rutschen
slipper clutch   Rutschkupplung f
slit   Schlitz m, Spalt m; Lamelle f *(Reifen)*
slot   Nut f, Schlitz m, Spalt m
key slot   Keilnut f
slotted   geschlitzt, Schlitz-
slotted bolt   Schlitzschraube f
slotted nut   Schlitzmutter f
slotted piston   Schlitzmantelkolben m
slot-type piston   Schlitzmantelkolben m
slow-running jet   Übergangskraftstoffdüse f, Übergangsbohrung f, Leerlaufdüse f
sludge   Schlamm m, Matsch m
sludge formation   Schlammbildung f
small end   Pleuelkopf m, Pleuelstangenkopf m
smoke/to   rauchen, qualmen
smoke   Rauch m, Qualm m
smoke emission   Rauchzahl f, Rauchdichte f, Rauchemission f
smoke limit   Rauchgrenze f
smoke limiter   Rauchbegrenzer m
smoke particulates   Rußteilchen f, Rußpartikel f

smooth running Rundlauf m, runder Motorlauf m
snap-on cap Schnappverschluß m
snap ring Sprengring m, Sicherungsring m
snow Schnee m
  snow chain Schneekette f
  snow chain ring Schneekettenring m
  snow tire Winterreifen m, Schneereifen m
snubber Reibungsstoßdämpfer m
socket Buchse f, Fassung f, Dose f, Sockel m; Lampenfassung f
  socket wrench Steckschlüssel m
  ball-and-socket gear change Kugelschaltung f
  ball socket Kugelbolzenpfanne f, Kugelkopfpfanne f, Kugelpfanne f
  cable socket Kabelschuh m
  connector socket Anhängersteckdose f, Kupplungssteckdose f
  female socket Steckbuchse f
  ignition circuit socket Zündleitungsstecker m
  lamp socket Glühlampenfassung f, Lampensockel m
  male socket Stecker m
  pipe socket Rohrstutzen m
  plug socket Steckdose f, Steckfassung f
  plug-in socket Steckfassung f, Steckdose f
  plug-receiving socket Kerzenaufnahmestutzen m
  round plug socket Rundsteckhülse f
  screening socket Entstörstecker m
  screw-in socket Einschraubstutzen m
  spark plug socket Zündkerzenstecker m
  suppressing socket Entstörstecker m
soft weich
  soft top Klappverdeck n, Faltdach n
  soft trim Verkleidung f, Auskleidung f
solder/to löten, weichlöten
soldered pipe Lötstutzen m
solenoid Magnetspule f
  solenoid switch Anlaßmagnetschalter m, Magnetschalter m
solid starr, fest
  solid axle Starrachse f
  solid injection direkte *(kompressorlose)* Einspritzung f
  solid injection engine Motor m mit mechanischer Einspritzung,
Motor m mit Direkteinspritzung, Direkteinspritzer m

solid rubber tire Vollgummireifen m, Vollreifen m
solid-state ignition system Transistor-Zündanlage f
solid tire Vollgummireifen m, Vollreifen m
solid wheel Scheibenrad n
soot/to rußen
soot Ruß m
 soot formation Rußbildung f
 soot-free combustion rußfreie Verbrennung f
sound Schall m
 sound absorbing material schalldämmendes Material n
 sound absorption Schallabsorption f, Schalldämmung f,
 Schalldämpfung f
 sound-damping plate Schalldämpferplatte f
 sound deadener Antidröhnmittel n, Antidröhnmasse f
 sound deadening Schalldämmung f
 sound deadening material schalldämmendes Material n,
 Antidröhnmittel n
 sound insulation mat Schalldämm-Matte f
 sound projector Schalltrichter m
 sound proofing Schalldämmung f
 sound proofing material schalldämpfendes Material n,
 schalldämmendes Material n
 sound reduction Schalldämpfung f
 sound volume Lautstärke f
space Raum m
 combustion space Verbrennungsraum m
 luggage space Kofferraum m
spacer Distanzstück n, Beilagscheibe f
 spacer bush Distanzbüchse f
 spacer disk Distanzscheibe f
 spacer piece Distanzstück n
 spacer ring Distanzring m
spanner Schraubenschlüssel m
 fork spanner Gabelschlüssel m
 hexagon spanner Sechskantschlüssel m
 hollow screw spanner Imbusschlüssel m
 ring spanner Ringschlüssel m
 wheel nut spanner Radmutterschlüssel m
spare Ersatz-, Reserve-

spare part Ersatzteil n
spare tire Ersatzreifen m
spare wheel Reserverad n, Ersatzrad n
spare wheel pan Ersatzradmulde f, Reserveradmulde f
spare wheel well Ersatzradmulde f, Reserveradmulde f
spark Funke m, Zündfunke m
spark adjustment Zündverstellung f
spark advance Vorzündung f, Frühzündung f
spark advance device Zündversteller m
spark advance/retard unit Zündversteller m
spark control Zündverstellung f, Zündzeitpunktverstellung f
spark discharge Funkenüberschlag m
spark electrodes Zündkerzenelektroden fpl
spark gap Elektrodenabstand m, Funkenstrecke f, Zündabstand m
spark ignition Fremdzündung f, Funkenzündung f
spark ignition engine Ottomotor m
spark knock Zündungsklopfen n
spark lag Zündverzug m
spark-over Funkenüberschlag m
spark-over voltage Überschlagspannung f
spark plug Zündkerze f
spark plug air gap Elektrodenabstand m
spark plug barrel Kerzengehäuse n
spark plug body Zündkerzenkörper m, Zündkerzengehäuse n,
Kerzengehäuse n
spark plug connector Zündkerzenstecker m, Kerzenstecker m
spark plug electrode Zündkerzenelektrode f
spark plug gap Elektrodenabstand n, Zündabstand m
spark plug insert Kerzeneinsatz m
spark plug insulator Zündkerzenstein m
spark plug lead Zündkerzenkabel n
spark plug opening Elektrodenabstand m, Zündabstand m
spark plug shell Zündkerzengehäuse n
spark-plug shield Funkenabschirmung f
spark plug socket Zündkerzenstecker m, Kerzenstecker m,
Zündkerzenanschlußstecker m, Kerzenanschlußstecker m
spark plug socket wrench Zündkerzenschlüssel m,
Kerzenschlüssel m
spark plug terminal Zündkerzenstecker m, Zündkerzenklemme f

spark plug tester Zündkerzenprüfgerät n
spark plug transfer hole Zündkanal m
spark plug wire Zündkerzenkabel n, Zündkabel n
spark retard Nachzündung f, Spätzündung f
spark setting Zündzeitpunkteinstellung f, Zündungseinstellung f,
Zündeinstellung f
spark timing Zündzeitpunktverstellung f, Zündverstellung f
central electrode of spark plug Kerzenstift m
centrifugal spark advance Fliehkraftverstellung f,
Fliehkraftversteller m, Fliehkraftzündversteller m
jump spark Überspringfunke m
vacuum spark advance Unterdruckzündverstellung f
sparking funkenbildend, Zünd-
  sparking plug Zündkerze f
  non-sparking funkenfrei
sparkless funkenfrei
speaker Lautsprecher m
  speaker balance Überblendregler m
special Sonder-, Spezial-
  special body Sonderaufbau m, Sonderkarosserie f,
  Spezialkarosserie f
  special purpose vehicle Sonderfahrzeug n, Spezialfahrzeug n
specific power output spezifische Leistung f, Hubraumleistung f,
Literleistung f
specified vorgeschrieben
  specified inflation pressure vorgeschriebener Reifendruck m
  specified tire pressure vorgeschriebener Reifendruck m
specification Vorschriften f
  test specification Prüfvorschrift f
speed Drehzahl f, Geschwindigkeit f, Gang m *(Getriebe)*
  speed-change gear Gangschaltung f, Schaltgetriebe n
  speed-dependent drehzahlabhängig
  speed governor Drehzahlregler m, Drehzahlbegrenzer m
  speed limit Höchstgeschwindigkeit f
  speed limiting device Drehzahlbegrenzer m
  speed range Drehzahlbereich m
  speed regulation Drehzahlregelung f
  change-speed gear Wechselgetriebe n
  driving speed Fahrgeschwindigkeit f

forward speed Vorwärtsgang m
high-speed hochtourig
low-speed nozzle kleine Düse f
maximum speed Höchstdrehzahl f, Höchstgeschwindigkeit f
normal speed Betriebsdrehzahl f
operating speed Betriebsdrehzahl f
rated speed Nenndrehzahl f
rotational speed Umlaufgeschwindigkeit f
starting speed Anlaßdrehzahl f, Startdrehzahl f
variable-speed governor Verstellregler m
speedometer Tachometer n, Geschwindigkeitsmesser m
speedometer reading Geschwindigkeitsanzeige f
record speedometer Fahrtenschreiber m
spherical kugelförmig, Raum-
spherical cam Raumnocken m
spherical cap Kalotte f, Kugelkalotte f
spigot Zapfen m
spigot trunnion Tragzapfen m
drive-shaft spigot (stub) Antriebswellenstumpf m
spike tire Spikereifen m
spill metering Überlaufdosierung f *(Einspritzpumpe)*
spill port Überlaufkanal m, Überströmkanal m
spin/to durchdrehen *(Räder)*
spindle Spindel f, Welle f, Druckstange f, Achswelle f
steering spindle Lenkspindel f
spiral Spirale f, Wendel f, Glühdraht m
spiral housing Spiralgehäuse n *(Turbolader)*
spiral spring Spiralfeder f, Schraubenfeder f
glow spiral Glühwendel f
heating spiral Glühspirale f
splash Spritz-, Schwall-
splash baffle Schwallblech n, Beruhigungstopf m, Schlingertopf m
splash guard Spritzschutz m, Spritzblech n
splash lubrication Spritzschmierung f, Schleuderschmierung f,
Spritzölschmierung f, Tauchschmierung f, Eintauchschmierung f
splash oil Spritzöl n
splash oiling Spritzschmierung f, Schleuderschmierung f,
Spritzölschmierung f, Tauchschmierung f, Eintauchschmierung f
splash plate (shield) Spritzblech n, Ablenkblech n

splash wall   Stirnwand f, Spritzwand f
splay   Sturz m, Radsturz m
spline/to   nuten, einkerben
splined   genutet, gekerbt
  splined hub   Kerbnabe f
  splined shaft   Keilwelle f, Nutwelle f
spline end   Schiebestück n
split axle   geteilte Achse f
split pin   Splint m
splitter gear   Zusatz-Schaltgruppe f, Zusatzgruppe f *(Getriebe)*
spoiler   Spoiler m, Luftleiteinrichtung f, Luftleitblech n
spoke   Speiche f
  spoke wheel   Speichenrad n
spontaneous ignition   Glühzündung f, Selbstzündung f
sports   Sport-
  sports car   Sportwagen m
  sports steering wheel   Sportlenkrad n
spot   Punkt-
  spot lamp   Suchscheinwerfer m
  spot welding   Punktschweißung f
spotlight   Suchscheinwerfer m
spray/to   besprühen; spritzlackieren
spray   Strahl m, Einspritzstrahl m
  spray angle   Spritzwinkel m, Sprühwinkel m
  spray atomization   Zerstäubung f *(Kraftstoffstrahl)*
  spray clutch   Freilauf m
  spray nozzle   Vergaserdüse f, Einspritzdüse f
  spray orifice   Düsenaustrittsöffnung f, Düsenöffnung f
  spray valve   Einspritzventil n
spraying nozzle   Spritzdüse f
spread/to   ausbreiten
spread-beam headlamp   Breitstrahler m
spreader   Expansionsring m *(Kupplung)*
spring   Feder f
  spring assembly   Federpaket n, Federsatz m
  spring balance   Federwaage f
  spring bearing   Federlager n
  spring bolt   Federbolzen m
  spring bracket   Federträger m, Federbock m

spring brake cylinder Federspeicherbremszylinder m
spring braking system Federspeicherbremsanlage f
spring buffer Federpuffer m
spring bush Federbüchse f
spring clamp Federbügel m
spring clip Federklemme f, Federschelle f, Federbügel m
spring deflection Federdurchbiegung f, Federweg m
spring disk Federscheibe f
spring eye Federauge n
spring hanger Federbock m
spring leaf Federblatt n
spring load Federbelastung f
spring-loaded federbelastet
spring mass gefederte Masse f
spring pad Federauflage f, Federunterlage f
spring plate Federblatt n, Federteller m
spring preload Federhärte f, Federvorspannung f
spring pressure Federdruck m
spring seat Federsitz m, Federteller m
spring shackle Federlasche f
spring steel Federstahl m
spring steel band Federstahlband n
spring steel insert Federstahleinlage f
spring strut Federbein n *(Radaufhängung)*
spring support Federträger m, Federstütze f
spring suspension Federaufhängung f
spring tension Federspannung f
spring travel Federweg m
spring washer Federring m, Federscheibe f, Federunterlage f
auxiliary spring Hilfsfeder f
clutch spring Kupplungsfeder f
coil spring Schraubenfeder f
compression spring Druckfeder f
contact spring Kontaktfeder f
disk spring Scheibenfeder f
flat spring Blattfeder f, Flachfeder f
front spring Vorderfeder f
laminated spring Blattfeder f
locking spring Haltefeder f

 nozzle spring Düsenfeder f
 pressure spring Druckfeder f
 rear spring Hinterfeder f
 reset spring Rückstellfeder f
 return spring Rückholfeder f
 safety spring Sicherungsfeder f
 shock absorbing spring Stoßfeder f
 spiral spring Schraubenfeder f, Spiralfeder f
 tension spring Zugfeder f
 torsion bar spring Drehstabfeder f
 valve spring Ventilfeder f
 valve spring disk Ventilfederteller m
sprinkle/to sprühen
sprocket Kettenrad n
 sprocket chain Zahnkette f
 sprocket wheel Kettenrad n
spur Stirn- *(Zahnrad)*
 spur gear Stirnrad n
 spur gear differential Stirnraddifferential n
 spur gearing Stirnradgetriebe n
 spur pinion Stirnkegelrad n, Stirnrad n
 spur wheel Stirnrad n
 coupling spur wheel Kupplungsstirnrad n
 differential spur gear Ausgleichstirnrad n, Differentialstirnrad n
squab Rückenlehne f, Stützkissen n
square rechteckig
 square bar Vierkantstab m
 square bolt Vierkantschraube f
 square head screw Vierkantschraube f
 square nut Vierkantmutter f
squareback Schrägheck n
squeal Quietschen n, Reifenquietschen n
squeegee Wischergummi m
stability Stabilität f
 cornering stability Kurvenfestigkeit f, Neigungsstabilität f
 side tilt stability Neigungsstabilität f
stabilizer Stabilisator m *(Radaufhängung)*; Querstabilisator m, Drehstab m
 stabilizer bar Stabilisator m, Querstabilisator m, Drehstab m

stabilizer bar suspension Stabilisatorlagerung f
stabilizer suspension Stabilisatorlagerung f
stage Stufe f
  double-stage zweistufig
  single-stage einstufig
staged carburetor Registervergaser m, Stufenvergaser m
stainless steel rostfreier Stahl m
stall/to abwürgen *(Motor)*
stamp Schlagstempel m, Stempel m
  stamp guide Stempelführung f
stamping Preßteil n, Stanzteil n
stand Ständer m
standard Norm f; serienmäßig, Standard-
  standard cycle Vergleichsprozess m
  standard design Normalausführung f
  standard drive Standardantrieb m, Hinterradantrieb m,
  Heckantrieb m
  standard equipment Grundausstattung f, serienmäßige
  Ausstattung f, Serienausstattung f, Standardausrüstung f
  standard fuel consumption Kraftstoff-Normverbrauch m
  standard model Standardmodell n, Grundmodell n
  standard part Normteil n, Serienteil n
  standard plug Meßkerze f
  standard rim Normfelge f, Meßfelge f
  standard size car Mittelklassenkraftfahrzeug n
  standard transmission Schaltgetriebe n, Handschaltgetriebe n
start/to anlassen
start Start m, Start-
  easy start jet Startdüse f
  false start Fehlstart m
startability Startfähigkeit f, Startfreudigkeit f, Startwilligkeit f,
  Startvermögen n
starter Anlasser m
  starter air jet Starterdüse f
  starter battery Anlaßbatterie f, Starterbatterie f, Startbatterie f
  starter button Anlaßdruckknopf m
  starter cable Anlasserkabel n
  starter control unit Betätigungseinrichtung f des Anlassers,
  Betätigungseinrichtung f des Starters, Anlaßschalter m,

Starterschalter m
starter interlock   Anlaßsperre f, Startverriegelung f
starter-lockout switch   Anlaßsperrschalter m
starter-motor speed   Anlasserdrehzahl f
starter-motor torque   Anlasserdrehmoment n
starter jet   Startdüse f
starter motor   Anlaßmotor m, Anlasser m, Anwurfmotor m
(Gasturbine)
starter pinion   Anlasserritzel n
starter protection   Anlasserschutz m
starter push-button   Anlaßschalter m, Anlasserdruckknopf m
starter ring gear   Anlasserzahnkranz m
starter switch   Anlaßschalter m
starter system   Anlaßanlage f, Anlaßsystem n, Anlasseranlage f
starter terminal   Anlasserklemme f
starter valve   Startventil n
centrifugal starter   Fliehkraftanlasser m
compressed-air starter   Druckluftanlasser m
engine starter   Anlasser m
gear shift starter   Schubtriebanlasser m
hot starter switch   Glühanlaßschalter m
ignition starter switch   Zündanlaßschalter m
screw shift pinion starter   Schubschraubtriebanlasser m
screw-push starter   Schubschraubtriebanlasser m
starting   Start-
starting ability   Starteigenschaft f, Startvermögen n
starting acceleration   Anfahrbeschleunigung f
starting aid   Starthilfe f, Starterleichterung f
starting air corrector jet   Starterluftdüse f
starting battery   Anlaßbatterie f
starting butterfly valve   Starterdrosselklappe f
starting carburetor   Startvergaser m
starting coil   Anlaßspule f
starting crank   Andrehkurbel f
starting device   Startanlage f, Anlaßanlage f
starting distance   Anfahrweg m, Anfahrstrecke f
starting jet   Startdüse f, Starterdüse f, Starterkraftstoffdüse f,
Starterbohrung f
starting magnetic switch   Anlaßmagnetschalter m

starting magneto   Anlaßmagnetzünder m
starting mechanism   Startvorrichtung f, Startvergaser m
starting mixture   Anlaßgemisch n
starting motor   Anlasser m, Starter m, Startermotor m
starting motor pinion   Anlasserritzel n, Starterritzel n
starting motor ring gear   Anlaßzahnkranz m, Starterzahnkranz m
starting nozzle   Starterdüse f
starting resistance   Anfahrwiderstand m
starting speed   Anlaßdrehzahl f, Startdrehzahl f
starting switch   Anlaßschalter m
starting torque   Anlaßdrehmoment n, Anlaufdrehmoment n,
Anzugsmoment n
starting valve   Startventil n
cold-starting aid   Kaltstarthilfe f
heater plug starting switch   Glühanlaßschalter m
heater starting switch   Glühanlaßschalter m
magneto starting switch   Magnetanlaßschalter m
static friction   Haftreibung f
static screen   Entstörkappe f
stationary gear   Ritzel n, Stirnrad n
station car   Kombiwagen m, Kombi m
stator   Stator m, Ständer m
stator frame   Ständergehäuse n
stay   Strebe f, Zuganker m
stay bolt   Abstützbolzen m
steady running   ruhiger Motorlauf m
steel   Stahl m
steel band   Stahlband n
all-steel body   Ganzstahlkarosserie f
cast steel wheel   Stahlgußrad n
spring steel   Federstahl m
steer/to   lenken, steuern
steering   Lenkung f, Lenk-
steering arm   Lenkhebel m, Spurstangenhebel m, Lenkspurhebel m
steering behavio[u]r   Lenkverhalten n
steering booster   Lenkhilfe f, Lenkunterstützung f
steering box   Lenkgetriebe n, Lenkgehäuse n,
Lenkgetriebegehäuse n
steering column   Lenksäule f

steering column change Lenkradschaltung f,
Lenkstockschaltung f
steering column gear shift Lenkradschaltung f,
Lenkstockschaltung f
steering column lock Lenkschloß n
steering column shift Lenkradschaltung f, Lenkstockschaltung f
steering column tube Mantelrohr n, Lenksäulenrohr n
steering damper Lenkungsdämpfer m
steering gear case Lenkgehäuse n
steering idler arm Lenkzwischenhebel m
steering knuckle Achsschenkel m
steering knuckle pin Achsschenkelbolzen m,
Achsschenkelzapfen m
steering lever Lenkhebel m
steering limiter Lenkbegrenzung f
steering offset Lenkrollradius m
steering pitman arm Lenkstockhebel m
steering rack Zahnstange f
steering rod Lenkstange f, Spurstange f
steering sector Lenksegment n
steering shaft Lenkwelle f, Lenkspindel f
steering shock absorber Lenkungsdämpfer m
steering spindle Lenkspindel f
steering tie rod Lenkspurstange f, Spurstange f
steering track rod Lenkspurstange f, Spurstange f
steering wheel Lenkrad n
steering wheel hub Lenkradnabe f
steering wheel turn Lenkradumdrehung f
steering worm Lenkschnecke f
front steering Vorderradlenkung f
hydraulically assisted steering Lenkung f mit hydraulischer
Lenkhilfe, hydraulische Servolenkung f
hydro steering gear Hydrolenkgetriebe n
left-hand steering Linkslenkung f
power-assisted steering Servolenkung f
power-steering mechanism Servolenkung f, Servolenkgetriebe n
rack-and-pinion steering gear Zahnstangenlenkgetriebe n
rack steering Zahnstangenlenkung f
recirculating-ball steering Kugelumlauflenkung f

worm-and-sector steering gear Schneckenlenkung f
stem Schaft m
  valve stem Ventilstange f, Ventilschaft m
step on the clutch pedal/to auskuppeln, auf die Kupplung treten
step Stufe f, Trittbrett n
  step cylinder Stufenzylinder m
  step-down Untersetzung f
  step-up Übersetzung f
  multi-step reduction gear Stufengetriebe n
stepped master cylinder Stufenhauptzylinder m
stick Stab m
  oil dip stick Ölmeßstab m
stick/to kleben, verkleben
sticking Festkleben n (Hängenbleiben) *(z.B. Kolbenringe)*
stiff suspension harte Federung f
stiffening plate Verstärkungsblech n, Verstärkungsplatte f,
  Versteifungsblech n
stock Materiallager n, Ersatzteillager n
  stock part Serienteil n
stone deflector Steinschlagschutz m
stoneguard Steinschlagschutz m
stop/to bremsen; anhalten
stop Anschlag m, Arretierung f, Sperre f
  stop angle Anschlagwinkel m
  stop buffer Anschlagpuffer m
  stop lamp Bremsleuchte f
  stop light Bremslicht n, Stopplicht n; Stoppleuchte f
  stop light switch Bremslichtschalter m
  stop pin Anschlagstift m
  stop valve Absperrventil n
  change-over stop Umschaltsperre f
  idle motion stop Leerlaufanschlag m
  pedal stop Pedalanschlag m
  rubber stop Gummianschlag m, Anschlaggummi m
stopping distance Bremsweg m, Bremsstrecke f
storage battery Akkumulator m, Akkumulatorbatterie f, Akku m
stoved enamel eingebrannter Lack m
stove-enamel finish Einbrennlackierung f
straight pin Zylinderstift m

strainer Sieb n, Filter n,m
strand Litze f
  strand wire Litzendraht m
strangler Drossel f, Drosselklappe f *(Vergaser)*
strap Band n, Riemen m
  cable strap Kabelband n
  earth (ground) strap Masseband n
striker Schließplatte f *(Tür)*
  striker pin Schließbolzen m
  striker plate Schließplatte f
strip/to abziehen, abnehmen
  strip the insulation/to abisolieren
strip Streifen m, Leiste f
  door weather strip Türdichtungsstreifen m
  fuse strip Schmelzstreifen m, Sicherungsstreifen m
stroboscope Stroboskop n
stroke Hub m, Kolbenhub m, Takt m
  ignition stroke Verbrennungshub m
  intake stroke Ansaughub m, Saughub m, Ladehub m
  no-load stroke Leer*lauf*hub m
  piston stroke Kolbenhub m
structure Anordnung f, Struktur f
  basic structure Fahrwerk n *(ohne Motor)*, Unterbau m
  reinforced belt structure Gürtelkonstruktion f *(Reifen)*
strut Strebe f
  axle strut Achsstrebe f
stub Stummel m, Stumpf m
  stub axle Achsschenkel m
  suction stub Saugstutzen m
stud Stehbolzen m, Stift m, Bolzen m
studded tire Spikereifen m
stuffing box Stopfbüchse f
style Ausführung f, Stil m
subcompact car PKW m der US-Kleinwagenklasse
suction Saugwirkung f
  suction cross-section Ansaugquerschnitt m
  suction filter Saugfilter n,m
  suction governor Unterdruckregler m
  suction nozzle Saugdüse f

suction pipe Ansaugrohr n, Saugrohr n, Saugleitung f
suction port Ansaugschlitz m
suction pump Absaugpumpe f
suction stroke Saughub m, Ansaughub m
suction stub Saugstutzen m
suction tube Saugrohr n
suction valve Ansaugventil n, Saugventil n
sump Ölfangschale f, Ölwanne f, Ölsumpf m
  oil sump Ölwanne f
  oil sump gasket Ölwannendichtung f
sun Sonne f
  sun gear Sonnenrad n
  sun roof Schiebedach n
  sun screen Sonnenblende f
  sun visor Sonnenblende f
  sun wheel Sonnenrad n
supercharge/to aufladen
supercharged engine aufgeladener Motor m, Kompressormotor m
supercharger Kompressor m, Lader m, Auflader m, Verdichter m
supercharging Aufladung f
  supercharging pressure Ladedruck m
  supercharging pressure adjusting device Ladedruckregler m
  supercharging pressure control Ladedruckregler m
  supercharging pressure ga[u]ge Ladedruckanzeige f,
  Ladedruckkontrolle f
  supercharging pressure limiter Ladedruckbegrenzer m
supplementary Zusatz-
  supplementary air Zusatzluft f
  supplementary air valve Zusatzluftventil n
supply Zuleitung f, Zufuhr f, Versorgung f
  supply pressure Förderdruck m
  supply pump Versorgungspumpe f, Förderpumpe f
  fuel-supply line Kraftstoffleitung f
support Abstützung f, Halter m, Stütze f, Träger m
  support bearing Stützlager n
  support tube Abstützrohr n, Stützrohr n
  mirror support Spiegelhalter m
  nozzle support Düsenhalter m
  rear spring support Hinterfederstütze f

supporting   Stütz-, Trag-, Halte-
  supporting angle   Auflagewinkel m, Tragwinkel m, Haltewinkel m
  supporting arm   Stützarm m, Tragarm m
  supporting axle   Tragachse f
  supporting frame   Stützrahmen m, Hilfsrahmen m
  supporting tube   Stützrohr n, Tragrohr n
suppress/to   entstören
suppressing socket   Entstörstecker m
suppression   Unterdrücken n
  interference suppression   Entstörung f
suppressor   Störschutzkondensator m, Drosselkondensator m
  suppressor cap   Entstörkappe f, Entstörstecker m *(Radio)*
  suppressor-type plug connector   Zündkerzenentstörstecker m
  interference suppressor   Entstörer m
surface   Oberfläche f
  surface appearance   Tragbild n
  surface crack   Oberflächenriß m
  surface ignition   Glühzündung f, Oberflächenzündung f
surge   Spannungsstoß m
suspension   Aufhängung f, Radaufhängung f, Federung f
  suspension alignment   Achsvermessung f, Achseinstellung f,
  Spureinstellung f
  suspension arm   Lenker m, Querlenker m
  suspension link   Lenker m, Achsstrebe f
  suspension link pivot   Lenkeranlenkung f
  suspension lower arm   unterer Lenker m, unterer Querlenker m
  suspension rate   Federkonstante f der Radaufhängung, Federrate f
  der Radaufhängung, Federerkennung f der Radaufhängung
  suspension system   Radaufhängung f
  suspension timing   Fahrwerksabstimmung f
  suspension tube   Tragrohr n, Achsrohr n
  engine suspension   Motoraufhängung f
  independent suspension   Einzelradaufhängung f
  rear suspension   Hinterachsaufhängung f
  wheel suspension   Radaufhängung f
sway   Seitenneigung f
  sway bar   Stabilisator m, Drehstab m
swept volume   Hubraum m, Hubvolumen n
swing axle   Schwingachse f, Pendelachse f

switch/to   schalten
  switch off/to   ausschalten, abschalten
  switch the main beam/to   aufblenden
switch   Schalter m
  switch box   Schalterdose f
  switch button   Schalterknopf m
  switch point   Schaltpunkt m, Schaltzeitpunkt m
  battery change-over switch   Batterieumschalter m
  battery main switch   Batteriehauptschalter m
  battery switch   Batterieumschalter m
  beam deflector switch   Abblendschalter m
  change-over switch   Umschalter m, Drehrichtungsschalter m
  contact switch   Kontaktschalter m
  dimming switch   Abblendschalter m
  dip switch   Abblendschalter m
  direction-indicating switch   Fahrtrichtungsanzeigeschalter m
  flasher indicator switch   Blinkschalter m
  foot dip switch   Fußabblendschalter m
  foot switch   Fußschalter m
  foot-operated dimming switch   Fußabblendschalter m
  foot-operated switch   Fußschalter m
  governor switch   Reglerschalter m
  heater starting switch   Glühanlaßschalter m
  hot starter switch   Glühanlaßschalter m
  ignition starter switch   Zündanlaßschalter m
  indicator switch   Fahrtrichtungsanzeigeschalter m,
Winkerschalter m
  inhibitor switch   Anlaßsperrschalter m
  limit switch   Endschalter m, Endausschalter m
  magnetic switch   Magnetschalter m
  magneto starting switch   Magnetanlaßschalter m
  main current switch   Hauptstromschalter m
  main switch   Hauptschalter m
  pedal-operated dip switch   Abblendfußschalter m
  pressure switch   Druckschalter m
  push button switch   Druckknopfschalter m
  rotary switch   Drehschalter m
  solenoid switch   Magnetschalter m
  starter switch   Anlaßschalter m

starting magnetic switch Anlaßmagnetschalter m
starting switch Anlaßschalter m
toggle switch Kippschalter m
switchboard Schaltbrett n, Schalttafel f
switchgear Betätigungseinrichtung f, Schaltanlage f
swivel Zapfen m, Drehpunkt m
swivel arm Schwenkarm m
swivel joint Kreuzgelenk n, Drehgelenk n
swivel pin Achsschenkelbolzen m, Achszapfen m, Drehzapfen m,
Lenkzapfen m
swivel window Drehfenster n, Dreiecksfenster n
swivel[l]ing radius Lenkrollradius m
synchromesh Synchronisierung f
synchromesh body Synchronkörper m
synchromesh gear Synchrongetriebe n, synchronisiertes Getriebe n
synchromesh mechanism Synchronisiereinrichtung f
synchromesh transmission synchronisiertes Getriebe n,
Synchrongetriebe n
synchronization Synchronisation f
synchronize/to synchronisieren
synchronized gearbox synchronisiertes Getriebe n,
Synchrongetriebe n, synchronisiertes Zahnradgetriebe n
synchronizing Synchronisier-, Synchron-
synchronizing ball Synchronkugel f
synchronizing lock Synchronriegel m
synchronizing cone Synchronkegel m
synchronizing disk Synchronscheibe f
synchronizing mechanism Synchronisiereinrichtung f
synchronizing slide collar Synchronschiebehülse f
synchronizing spring Synchronfeder f
synchronous synchron
synchro ring Synchronring m
synthetic bearing Kunststofflager n
system System n, Anlage f
exhaust system Abgasanlage f, Auspuffanlage f
fuel system Kraftstoffanlage f, Brennstoffanlage f *(Gasturbine)*
ignition system Zündanlage f
t-piece T-Stück n
table Tisch m, Tafel f, Platte f, Tabelle f

tachograph Tachograph m, Fahrtenschreiber m
tachometer Drehzahlmesser m, Tourenzähler m
tail Heck n
  tail flasher lamp Schluß-Blinkleuchte f
  tail gate Hecktür f, Heckklappe f
  tail lamp Schlußleuchte f, Rückleuchte f
  tail light Schlußlicht n, Rücklicht n
  tail light supervision Schlußlichtkontrolle f
  tail pipe Auspuffendrohr n
  flash-tail lamp Blink-Schlußleuchte f
  stop-tail lamp Brems-Schlußleuchte f
take-off Anfahren n
tandem axle Doppelachse f, Tandemachse f
tank Tank m, Behälter m
  tank cap Tankdeckel m, Tankverschluß m
  tank capacity Tankinhalt m
  tank ga[u]ge Vorratsgeber m
  tank transmitter Tankgeber m
  compensating tank Ausgleichbehälter m
  fuel tank Kraftstoffbehälter m
  preliminary air tank Vorluftbehälter m
  radiator tank Wasserkasten m
tanker Tankwagen m, Tankfahrzeug n
tap/to 1. anzapfen; 2. Gewinde n schneiden
tape Band n, Streifen m
  adhesive tape Klebeband n
  insulating tape Isolierband n
taper Verjüngung f, Kegel m
  taper base rim Schrägschulterfelge f
  taper bead seat rim Schrägschulterfelge f
  taper bolt Konusbolzen m
tapered konisch, kegelförmig
  tapered bead seat rim Schrägschulterfelge f
  tapered bearing Kegellager n
  tapered rim Schrägschulterfelge f, Steilschulterfelge f
  tapered roller bearing Kegelrollenlager n, Schrägrollenlager n
  tapered seat Schrägschulter f
tappet Stößel m, Ventilstößel m
  tappet clearance Ventilstößelspiel n, Ventilspiel n

tappet guide Stößelführung f
tappet roller Stößelrolle f
tappet wrench Ventilschlüssel m
roller tappet Rollenstößel m
valve tappet Ventilstößel m
telescopic Teleskop-
  telescopic antenna Teleskopantenne f
  telescopic joint Schiebegelenk n
  telescopic shock absorber Teleskopstoßdämpfer m
  telescopic tube Teleskoprohr n, Ausziehrohr n
tell-tale lamp Kontrollampe f, Anzeigeleuchte f
temperature Temperatur f
  temperature coefficient Temperaturkoeffizient m
  temperature ga[u]ge Temperaturmeßgerät n
  temperature indicator Temperaturanzeiger f
  temperature pick-up Temperaturgeber m
  temperature reading Temperaturanzeige f
  temperature sensor Temperaturfühler m
  temperature switch Temperaturschalter m
  operating temperature Betriebstemperatur f
tensile strength Zugfestigkeit f
tension Spannung f, Zug m *(meist mechanisch)*
  tension bush Spannbüchse f
  tension pulley Spannrolle f
  tension spring Zugfeder f, Spannfeder f
  tension strut Zugstab m, Zugstange f
  high tension Hochspannung f
  high-tension ignition cable Hochspannungszündleitung f
  low tension Niederspannung f
tensioner Spanneinrichtung f, Spanner m
  chain tensioner Kettenspanner m
tensioning lever Spannhebel m
terminal Anschlußklemme f, Klemme f; Pol m *(Batterie)*
  terminal lug Anschlußfahne f
  terminal nut Anschlußmutter f
  terminal voltage Klemmspannung f
  battery terminal Batterieklemme f
  connecting terminal Anschlußklemme f
  earthing (grounding) terminal Masseklemme f

test Prüfung f, Test m, Versuch m
  test bench Prüfstand m, Motorprüfstand m, Motorenprüfstand m
  test conditions Prüfbedingungen fpl, Testbedingungen fpl, Versuchsbedingungen fpl
  test cycle Prüfzyklus m, Testzyklus m
  test device 1. Prüfvorrichtung f, Prüfeinrichtung f; 2. Prüfpuppe f, Testpuppe f, Versuchspuppe f, Dummy m
  test equipment Prüfeinrichtung f
  test load Prüflast f, Prüfbelastung f
  test rig Prüfstand m, Prüfgestell n
  test run Prüflauf m, Probelauf m
  test site Prüfgelände n, Testgelände n, Versuchsgelände n
  test stand Prüfstand m
  capacity test Kapazitätsprüfung f
  corrosion test Korrosionsprüfung f
  load test Belastungsprüfung f
tester Prüfgerät n
  acid tester Säureprüfer m
  battery tester Batterieprüfer m, Batteriemeßinstrument n
testing instrument Prüfgerät n
textile upholstery Stoffpolsterung f
theft protection Diebstahlsicherung f, Diebstahlschutz m
thermal thermisch, Wärme-
  thermal expansion Wärmeausdehnung f
  thermal reactor thermischer Reaktor m
  thermal slot Kolbenschlitz m
thermocouple Thermoelement n
thermostat Temperaturregler m, Thermostat m
  thermostat body Thermostatgehäuse n
  thermostat housing Thermostatgehäuse n
thermostatic spring Thermostatfeder f, Bimetallfeder f
thermoswitch Thermoschalter m
thinner Verdünner m
thread Gewinde n
  thread fit Gewindepassung f
  metric thread metrisches Gewinde n
threaded Gewinde-
  threaded bolt Gewindebolzen m
  threaded bush Gewindebüchse f

threaded insert   Gewindeeinsatz m
threaded piece   Gewindestück n
threaded pin   Gewindestift m
threaded pivot   Gewindezapfen m
three Drei-
  three-dimensional cam   Raumnocken m, dreidimensionaler
  Nocken m
  three-phase alternator   Drehstromgenerator m
  three-point attachment   Dreipunktbefestigung f
  three-point belt   Dreipunktgurt m
throat   Lufttrichter m, Vergaserlufttrichter m
throttle/to   drosseln
throttle   Drossel f, Drosselklappe f, Vergaserdrossel f
  throttle adjusting screw   Leerlaufeinstellschraube f,
  Leerlaufbegrenzungsschraube f
  throttle blade   Drosselklappe f
  throttle body   Drosselklappengehäuse n, Drosselklappenteil n
  throttle bore   Drosselbohrung f
  throttle carrier   Klappenstutzen m
  throttle control   Drosselklappensteuerung f, Klappenregler m
  throttle control cable   Gasseilzug m
  throttle control lever   Drosselklappenhebel m, Drosselhebel m
  throttle housing   Drosselklappenstutzen m, Klappenstutzen m
  throttle lever   Drosselhebel m, Drosselklappenhebel m
  throttle linkage   Drosselklappengestänge n, Gasgestänge n
  throttle opening angle   Drosselklappenöffnungswinkel m
  throttle orifice   Drosselbohrung f
  throttle shaft   Drosselklappenwelle f
  throttle slide   Drosselschieber m, Flachschieber m
  throttle spindle   Drosselklappenwelle f
  throttle switch   Drosselklappenschalter m
  throttle valve   Drosselklappe f
  throttle valve shaft   Drosselklappenwelle f
  choke throttle lever   Handgashebel m
  full throttle   Vollgas n
throttling loss   Drosselverlust m
throw the gear into reverse/to   den Rückwärtsgang einlegen
thrust   Schub m, Axialschub m, Axialdruck m
  thrust bearing   Drucklager m, Axialdrucklager n, Axiallager n

thrust release bearing Kupplungsausrücklager n, Ausrücklager n
thrust ring Anpreßring m
thrust rod Schubstange f, Schubstrebe f
thrust washer Druckscheibe f
thyristor Thyristor m
tie rod Spurstange f, Lenkspurstange f
tightener Spanner m
  tightener pulley Spannrolle f, Spannrad n
  belt tightener Riemenspanner m
  chain tightener Kettenspanner m
tightening torque Anzugsdrehmoment n, Anziehdrehmoment n
tightening wheel Spannrad n
tilt Neigung f, Schrägstellung f
  tilt adjustment Neigungsverstellung f
time Zeit f, Dauer f, Zeitdauer f
  time switch Zeitschalter m
  discharge time Entladezeit f
  reaction time Ansprechzeit f, Reaktionszeit f
  starting time Anlaufzeit f, Anlaßzeit f
timer Taktgeber m
  injection timer Spritzversteller m
  ignition timer Zündverteiler m
  ignition timer distributor Zündverstellverteiler m
timing Zeitvorgabe f, Zeiteinstellung f
  timing angle Verstellwinkel m, Zündverstellbereich m
  timing case cover Steuergehäusedeckel m
  timing chain Steuerkette f, Nockenwellenantriebskette f
  timing control Zündverstellung f
  timing cover Steuergehäusedeckel m
  timing device Zündversteller m
  timing gear Nockenwellenantrieb m, Nockenwellenantriebsrad n; Ventilsteuerung f
  timing gear case Steuergehäuse n, Nockenwellenantriebsgehäuse n
  timing mark Totpunktmarke f
  timing range Verstellbereich m *(Zündung)*
  camshaft timing gear Nockenwellenantriebsrad n
  centrifugal timing device Fliehkraftversteller m
  centrifugally operated timing device Fliehkraftversteller m
  flywheel timing mark Schwungradmarke f

ignition timing   Zündverstellung f, Zündzeitpunktverstellung f
injection timing   Einspritzverstellung f *(Dieselmotor)*
spark timing   Zündmomenteinstellung f
vacuum timing control   Unterdruckzündverstellung f
tinted window   getönte Scheibe f
tipper   Kipper m
tipping gear   Kippvorrichtung f
tire   Reifen m
  tire adhesion   Reifenhaftung f
  tire bead   Reifenwulst m, Wulst m
  tire bead separation   Reifenwulstablösung f, Wulstablösung f
  tire body   Reifenkarkasse f, Karkasse f, Reifenunterbau m
  tire canvas   Reifengewebe n
  tire chafing strip   Felgenband n
  tire chain   Radkette f, Schneekette f, Gleitschutzkette f
  tire contact area   Reifenaufstandsfläche f, Aufstandsfläche f, Latsch m
  tire cord   Reifengewebe n
  tire cornering stiffness   Kurvensteifigkeit f, Seitenkraftbeiwert m
  tire ga[u]ge   Reifendruckprüfer m, Luftdruckprüfer m
  tire grip   Reifenhaftung f, Reifengriffigkeit f, Haftung f, Griffigkeit f, Bodenhaftung f
  tire groove   Profilrille f
  tire inflation pressure   Reifenluftdruck m, Reifendruck m
  tire iron   Reifenmontierhebel m
  tire pressure   Reifenluftdruck m, Reifendruck m
  tire pressure ga[u]ge   Reifendruckprüfer m, Luftdruckprüfer m
  tire side wall   Reifenflanke f, Seitenwand f
  tire tread   Reifenlauffläche f, Lauffläche f, Protektor m
  tire tread width   Spurweite f, Spur f
  solid-rubber tire   Vollgummi-Reifen m
toe   Spur f, Spurwinkel m, Radspur f, Radspurwinkel m
  toe-in   Vorspur f
  toe-out   Nachspur f
toggle lever   Nockenhebel m, Bremsnockenhebel m
toggle switch   Kippschalter m
tool   Werkzeug n
  tool kit   Werkzeugkasten m
  tool set   Werkzeugsatz n

tooth pitch Zahnteilung f
thoothed gezahnt, Zahn-
  toothed belt Zahnriemen m
  toothed chain Zahnkette f
  toothed gearing Zahnradgetriebe n
  toothed pinion Zahnritzel n
  toothed rim Zahnkranz m
  toothed washer Zahnscheibe f
top Dach n *(Karosserie)*
  top dead center oberer Totpunkt m
  top gear höchster (schnellster) Gang m
  top-up quantity Nachfüllmenge f
  collapsible top Faltdach n, Klappverdeck n
  folding top Klappverdeck n, Verdeck n
  roof top carrier Dachgepäckträger m
torque Drehmoment n, Torsionskraft f, Verdrehkraft f
  torque arm Schubstange f
  torque converter Drehmomentwandler m
  torque rod Schubstange f
  torque wrench Drehmomentschlüssel m
  rated torque Nenndrehmoment n
  self-aligning torque Rückstellmoment n *(Lenkung)*
  starting torque Anlaßdrehmoment n, Anlaufdrehmoment n,
  Anzugsmoment n
torsion Torsion f, Verdrehung f, Verwindung f
  torsion bar Drehstab m, Torsionsfeder f, Drehstabfeder f
  torsion bar spring Drehstabfeder f
  torsion bar stabilizer Drehstab-Stabilisator m
  torsion bar suspension Drehstabfederung f
  torsion resistance Verwindungssteifigkeit f, Verdrehsteifigkeit f
  torsion spring Torsionsfeder f
  torsion strength Torsionsfestigkeit f, Verdrehfestigkeit f
total weight Gesamtgewicht n
touch-up stick Lackstift m
tow/to abschleppen
tow Schlepp-, Abschlepp-
  tow bar Abschleppstange f, Schleppstange f, Zugstange f
  tow-bar coupling Abschleppkupplung f
  tow coupling Anhängerkupplung f, Anhängevorrichtung f

tow rod   Abschleppstange f
tow rope   Abschleppseil n
tow start   Anschleppen n, Schleppstart m
tow truck   Abschleppfahrzeug n
towing   Schleppen n
  towing device   Zugvorrichtung f, Anhängevorrichtung f,
  Anhängekupplung f, *(mechanische)* Kupplungseinrichtung f
  towing lug   Abschleppöse f
  towing rod   Abschleppstange f
  towing rope   Abschleppseil n
  towing service   Abschleppdienst m
  towing vehicle   Abschleppfahrzeug n, abschleppendes Fahrzeug n,
  Zugfahrzeug n
track   Spurweite f, Spur f, Radspur f
  track alignment   Radausfluchtung f
  track control arm   Querlenker m
  track groove   Spurrille f, Spurrinne f
  track rod   Spurstange f, Lenkspurstange f
  track rod arm   Spurstangenhebel m, Lenkspurhebel m
  track width   Spurbreite f, Spurweite f, Spur f
traction   Traktion f, Zugkraft f, Zugvermögen n
  traction gear   Geländegang m
  traction power   Durchzugsvermögen n, Zugkraft f,
  Durchzugskraft f, Zugleistung f
tractive   Zug-
  tractive force   Zugkraft f, Triebkraft f, Motorzugkraft f
  tractive power   Durchzugsvermögen n, Zugkraft f,
  Durchzugskraft f, Zugleistung f
tractor   Zugfahrzeug n, Zugmaschine f, Zugwagen m, Traktor m,
  Schlepper m, Sattelschlepperzugmaschine f,
  Sattelschlepperzugfahrzeug n
  tractor brake valve   Zugwagen-Bremsventil n
  tractor-trailer unit   LKW-Zug m, Lastzug m,
  Sattelschlepperzug m
traffic   Verkehr m
  traffic congestion   Verkehrsstockung f
  traffic density   Verkehrdichte f
  traffic flow   Verkehrsfluß m
  traffic intensity   Verkehrsdichte f

traffic jam Verkehrsstockung f, Verstopfung f
traffic junction Verkehrsknotenpunkt m, Kreuzung f
traffic light Verkehrsampel f, Ampel f
traffic signal n Verkehrsignal n
trafficator Winker m, Fahrtrichtungsanzeiger m
trail Nachlauf m *(Vorderräder)*
trailer Anhänger m, Fahrzeuganhänger m
  trailer brake cylinder Anhängerbremszylinder m
  trailer coach Omnibusanhänger m
  trailer coupling Anhängerkupplung f
  trailer load Anhängelast f
  trailer socket Anhängersteckdose f
  trailer towing device Anhängerkupplung f, Anhängevorrichtung f am Anhänger
  trailer truck Lastzug m
trailing Schlepp-, Nachlauf-
  trailing arm Trapezlenker m, Längslenker m
  trailing axle Schleppachse f, Nachlaufachse f
  trailing link Längslenker m
  trailing shoe Ablaufbacke f
  trailing strut Längslenker m
transaxle Achseinheit f *(mit Getriebe , Kupplung f und Differential)*
transfer box Verteilergetriebe n, Vorschaltgetriebe n
transformer Transformator m
transistor Transistor m
  transistor flashing unit Transistorblinkgeber m
  transistor ignition Transistorzündung f
transmission Übersetzung f, Getriebe n, Kraftübertragung f
  transmission bell housing Kupplungsglocke f
  transmission case Getriebegehäuse n
  transmission case flange Getriebeflansch m
  transmission chain Transmissionskette f
  transmission control Schalthebel m, Schaltknüppel m, Getriebewählhebel m, Wählhebel m
  transmission cover Getriebedeckel m
  transmission housing Getriebegehäuse n
  transmission main shaft Getriebehauptwelle f, Hauptwelle f
  transmission oil Getriebeöl n
  transmission oil cooler Getriebeölkühler m

transmission output shaft   Getriebeabtriebswelle f,
Getriebeausgangswelle f
transmission ratio   Übersetzungsverhältnis n
transmission ratios   Getriebesprünge mpl, Getriebeabstufung f
transmission selector control   Getriebewählhebel m,
Wählhebel m
transmission selector indicator   Wählhebelstellungsanzeige f,
Ganganzeige f
transmission selector lever   Getriebewählhebel m, Wählhebel m
transmission shaft   Antriebswelle f, Getriebeeingangswelle f
transmission tunnel   Getriebetunnel m, Kardantunnel m,
Mitteltunnel m
auxiliary transmission   Zusatzgetriebe n
power transmission   Kraftübertragung f
shiftless transmission   automatisches Getriebe n
transmitter   Geber m, Sender m
signal transmitter   Signalgeber m
tank transmitter   Tankgeber m
transverse   quer
transverse engine   quer eingebauter Motor m, Quermotor m
transverse link   Querlenker m
transverse member   Querträger m, Rahmenquerträger m,
Querschweller m, Verstrebung f
transverse rib   Querrippe f
travel   Wegstrecke f; Fahrt f
travel direction   Fahrtrichtung f
dead travel   Leerweg m, toter Gang m
direction of travel   Fahrtrichtung f
tray   Ablagefach n, Schale f
tread   Lauffläche f, Reifenlauffläche f, Protektor m; Spurweite f,
Spur f, Radspur f
tread bar   Profilstollen m, Laufflächenstollen m
tread bracing   Laufflächenverstärkung f
tread compound   Laufflächenmischung f
tread contact   Reifenaufstandsfläche f, Aufstandsfläche f, Latsch m
tread depth   Profiltiefe f, Laufflächentiefe f
tread depth ga[u]ge   Profiltiefenmesser m
tread design   Laufflächenprofil n
tread element   Profilstollen m

tread groove   Profilrille f
tread pattern   Laufflächenprofil n, Profilmuster n,
Laufflächengestaltung f
tread pattern depth   Profiltiefe f, Laufflächentiefe f
tread rib   Profilrippe f, Laufflächenrippe f
tread separation   Laufflächenablösung f
tread shoulder   Schulter f
tread wear   Laufflächenabnutzung f, Laufflächenverschleiß m,
Reifenverschleiß m, Reifenabnutzung f
tread width   Laufflächenbreite f; Spurweite f, Spur f
(Radaufhängung)
treatment   Bearbeitung f, Veredelung f
trial   Probe f, Versuch m
triangular reflector   Dreieckrückstrahler m
trigger   Auslöser m
trim   Zierteile npl, Karosserieverzierung f
  trim panel   Verkleidung f, Innenverkleidung f, Verkleidungstafel f
  trim ring   Zierring m; Abdeckblende f *(Scheinwerfer)*
trimplate   Blende f, Abdeckblende f
t-ring   T-Ring m
trip   Reise f, Fahrt f
  trip meter   Tageskilometerzähler m
triple-tone horn   Dreiklanghorn n
triplex glass   Verbundglas n
tropic-proof   tropenfest
truck   Lastkraftwagen m
  truck cab   LKW-Fahrerhaus n
  truck train   Lastzug m
true running   Rundlauf m
trumpet   Fanfare f, Schalltrichter m
trunk   Kofferraum m, Gepäckraum m
  trunk capacity   Kofferraumfassungsvermögen n,
  Kofferraumvolumen n
  trunk carpeting   Kofferraumauskleidung f
  trunk floor   Kofferraumboden m, Kofferboden m
  trunk lid   Kofferraumdeckel m
  trunk lid support   Kofferraumdeckelstütze f
  trunk light   Kofferraumbeleuchtung f
  trunk partition panel (wall)   Kofferraumtrennwand f

trussing Unterzug m
tube Rohr n, Schlauch m, Luftschlauch m
  tube bend Rohrbogen m
  tube elbow Rohrkrümmer m
  air inlet tube Lufteintrittstutzen m
  air intake tube Luftansaugrohr n
  angle tube (pipe) Winkelrohrstutzen m
  axle tube Achsrohr n
  choke tube Drosselleitung f, Drosselkanal m, Lufttrichter m,
  Mischrohr n *(Vergaser)*
  compressed-air tube Druckluftschlauch m
  corrugated tube Wellrohr n
  damping tube Dämpferrohr n
  exhaust tube Austrittsrohr n, Auspuffrohr n
  filling tube Einfüllrohr n
  flexible tube Schlauch m
  guide tube Führungsrohr n
  inner tube Luftschlauch m *(Reifen)*
  overflow tube Überlaufrohr n
  pressure tube Druckrohr n
  telescopic tube Ausziehrohr n, Teleskoprohr n
  venturi tube Venturirohr n
tubeless schlauchlos
  tubeless tire schlauchloser Reifen m
tubular rohrförmig; Rohr-
  tubular axle Rohrachse f
  tubular construction Rohrrahmenkonstruktion f
  tubular frame Rohrrahmen m
  tubular radiator Röhrenkühler m
  tubular ribbed radiator Röhren-Lamellen-Kühler m
  tubular shaft Gelenkrohrwelle f
tumbler switch Kippschalter m
tunnel Tunnel m
  transmission gear tunnel Getriebetunnel m
turbine Turbine f, Abgasturbine f
  turbine blade Turbinenschaufel f
  turbine rotor Turbinenrotor m
  turbine vane Turbinenschaufel f
  turbine wheel Laufrad n, Turbinenrad n

turbocharged engine Motor m mit Abgasturbolader, Turbomotor m
turbocharger Turbolader m
  turbocharger compressor Turboladerverdichter m, Turboladergebläse n
  turbocharger engine Motor m mit Abgasturbolader, Turbomotor m
  turbocharger lag Turboladerverzug m, Ansprechverzögerung f des Turboladers
turbosupercharger Abgasturbolader m
turbulence chamber Wirbelkammer f
turn/to abbiegen, wenden
  turn out/to ausschalten
turn Windung f, Umdrehung f
  turn indicator Fahrtrichtungsanzeiger m, Blinker m
  turn indicator lamp Fahrtrichtungsanzeigeleuchte f, Blinkerleuchte f, Blinkleuchte f
  turn lever Laufhebel m
  turn signal Blinker m
  turn signal flasher Blinkgeber m, Blinkrelais n
  turn signal indicator Fahrtrichtungsanzeigerkontrolle f, Blinkerkontrolle f
  turn signal lamp Fahrtrichtungsanzeigeleuchte f, Blinkerleuchte f, Blinkleuchte f, Blinker m
  turn signal lever Blinkerhebel m
  turn signal switch Blinkerschalter m
  turn signal tell-tale Fahrtrichtungsanzeigerkontrolleuchte f, Blinkerkontrolleuchte f
  turn switch Drehschalter m
  left-hand turn Linksabbiegen n
  right-hand turn Rechtsabbiegen n
turning Wende-
  turning circle Wendekreis m, Spurkreis m
  turning circle diameter Wendekreisdurchmesser m, Spurkreisdurchmesser m, kleinster Spurkreisdurchmesser m
  turning clearance circle Wendekreis m
  turning knob Drehknopf m
twin Doppel-
  twin-bar bumper Doppelstoßfänger m
  twin-branch exhaust system Zweirohr-Auspuffanlage f,

Doppelrohr-Auspuffanlage f, Doppelauspuffanlage f
twin choke carburetor Doppelvergaser m
twin control arm suspension Doppelquerlenkeraufhängung f
twin exhaust pipe Doppelauspuffrohr n
twin headlamps Doppelscheinwerfer mpl
twin-rotor engine Zweischeibenmotor m, Zweifachmotor m
twin tire Zwillingsreifen m, Doppelreifen m
twin-wishbone suspension Doppelquerlenkeraufhängung f
twist Verdrehung f, Torsion f
  twist spring Drehfeder f, Torsionsfeder f
  groove twist drill Senker m
two zwei
  two-phase carburetor Registervergaser m
  two-piece rim zweiteilige Felge f
  two-pole zweipolig
  two-speed zweistufig, zweigängig
  two-speed blower zweistufiges Gebläse n, Zweistufengebläse n,
  zweistufiger Ventilator m
  two-stage Zweistufen-, zweistufig
  two-stage carburetor Registervergaser m
  two-stroke diesel engine Zweitakt-Dieselmotor m
  two-stroke engine Zweitaktmotor m
  two stroke engine Zweitaktmotor m
  two stroke mixture Zweitaktgemisch n, Gemisch n
  two-way cock (tap) Zweiwegehahn m
  two-way valve Zweiwegventil n
  two-wheel trailer Einachsanhänger m
tyre s.tire
u-bolt Federbügel m
u-rail U-Schiene
unbalance Unwucht f
unbalanced unausgewuchtet, nicht ausgewuchtet
unbelted unangeschnallt, nicht angeschnallt
underbody Unterseite f des Aufbaus, Bodengruppe f, Fahrgestell n
  underbody coating Unterbodenschutz m
  underbody protection Unterbodenschutz m
undercarriage Fahrwerk n, Unterbau m, Fahrgestell n
undercoat Unterbodenschutz m
underfloor engine Unterflurmotor m

underseal Unterbodenschutz m
understeer/to untersteuern
understeering performance untersteuerndes Eigenlenkverhalten n
uniflow scavenging Gleichstromspülung f
union nut Anschlußmutter f, Überwurfmutter f
unit Einheit f, Aggregat n
  unit body selbsttragende Karosserie f
  unit construction body selbsttragende Karosserie f
  flasher unit Blinkgeber m
  heating unit Heizelement n
  overdrive unit Schnellganggetriebe n
  power per unit of displacement Hubraumleistung f,
  Literleistung f
  radio-shielding unit Entstörer m
  vacuum advance unit Frühdose f
  vacuum unit Unterdruckversteller m
unitized body selbsttragende Karosserie f
universal Universal-
  universal drive Kardanantrieb m
  universal joint Kreuzgelenk n, Kardangelenk n, Antriebsgelenk n,
  Universalgelenk n, Doppelgelenk n
  universal joint yoke Kreuzgelenkgabel f
  universal shaft drive Kardanantrieb m
unleaded unverbleit
  unleaded fuel unverbleiter Kraftstoff m, bleifreier Kraftstoff m
  unleaded gasoline unverbleites Benzin n, bleifreies Benzin n
unloaded weight Leergewicht n
unscrew/to abschrauben
unseating Abdrücken n des Wulstes vom Felgenhorn
updraft carburetor Steigstrom-Vergaser m
upholstery Polsterung f, Sitzpolsterung f, Sitzpolster n
upper Ober-
  upper beam Fernlicht n
  upper torso strap Oberkörpergurt m, Schultergurt m
upshift/to hochschalten, hinaufschalten
u-turn Kehrtwendung f, Wendung f
V-belt Keilriemen m
  V-belt pulley Keilriemenrolle f, Keilriemenscheibe f
  V-cylinder block Zylinderanordnung f in V-Form

narrow **V-belt**   Schmalkeilriemen m
vacuum   Vakuum n, Unterdruck m
  vacuum advance mechanism   Unterdruckzündversteller m
  vacuum box   Unterdruckdose f
  vacuum control   Unterdrucksteuerung f
  vacuum-controlled carburetor   Vergaser m mit
  Unterdrucksteuerung
  vacuum-controlled spark timing   Unterdruckzündverstellung f
  vacuum distributor   Unterdruckverteiler m
  vacuum fuel pump   Unterdruckförderpumpe f
  vacuum governor   Unterdruckregler m
  vacuum ignition adjustment   Unterdruckzündverstellung f
  vacuum pipe   Unterdruckleitung f, Unterdruckrohr n
  vacuum pump   Vakuumpumpe f, Unterdruckpumpe f
  vacuum spark advance mechanism   Unterdruck-
  zündverstelleinrichtung f
  vacuum timing control   Unterdruckzünd*zeitpunkt*verstellung f
  vacuum timing unit   Unterdruckversteller m
value   Wert m, Zahlenwert m
  heat value   Wärmewert m
  nominal value   Nennwert m
  saturation value   Sättigungswert m
  set value   Sollwert m
valve   Ventil n
  valve actuating mechanism   Ventiltrieb m
  valve adjustment   Ventileinstellung f
  valve ball   Ventilkugel f
  valve body   Ventilkörper m, Ventilgehäuse n
  valve cap   Ventilkappe f, Staubkappe f, Ventilstaubkappe f,
  Ventilaufsatz m, Ventilverschraubung f
  valve clearance   Ventilspiel n
  valve clearance ga[u]ge   Ventilspiellehre f
  valve collar   Ventilring m
  valve cone   Ventilkegel m
  valve cover   Ventildeckel m
  valve disk   Ventilteller m, Ventilscheibe f
  valve face   Ventilsitzfläche f
  valve flap   Ventilklappe f
  valve gear   Ventiltrieb m, Ventilantrieb m, Ventilmechanismus m,

Ventilsteuerung f
valve grinding   Ventileinschleifen n
valve guide   Ventilführung f
valve head   Ventilteller m
valve housing   Ventilgehäuse n
valve housing cover   Ventilgehäusedeckel m
valve inlet   Ventileinlaß m
valve insert   Ventileinsatz m
valve lever   Schwinghebel m, Kipphebel m, Schlepphebel m
valve lift   Ventilhub m, Ventilanhebung f
valve lifter   Ventilstößel m, Stößel m
valve lifting velocity   Ventilöffnungsgeschwindigkeit f
valve mechanism   Ventiltrieb m, Ventilantrieb m,
Ventilmechanismus m, Motorsteuerung f
valve opening   Ventilhub m, Ventilöffnung f
valve outlet   Ventilauslaß m
valve overlap   Ventilüberschneidung f
valve play   Ventilspiel n
valve port   Ventilkanal m
valve push rod   Ventilstößelstange f, Stößelstange f,
Ventilstoßstange f, Stoßstange f
valve reseating   Ventileinschleifen n
valve ring   Ventilring m
valve rocker   Kipphebel m, Schwinghebel m
valve seat   Ventilsitz m
valve setting   Ventileinstellung f
valve shank   Ventilschaft m
valve spring   Ventilfeder f
valve spring cotter   Ventilkeil m
valve spring disk   Ventilfederteller m
valve spring retainer   Ventilfederteller m
valve stem   Ventilstange f, Ventilschaft m
valve stem sealing   Ventilschaftabdichtung f
valve stroke   Ventilhub m
valve support   Ventilträger m
valve tappet   Ventilstößel m
valve throat   Ventildurchlaß m
valve timing   Steuerzeiteinstellung f *(Ventil)*
valve timing diagram   Steuerdiagramm n

valve train   Ventilantrieb m, Ventilmechanismus m, Motorsteuerung f
valve wrench   Ventileinstellschlüssel m, Ventilschlüssel m
air vent valve   Entlüfterventil n
auxiliary air valve   Zusatzluftregler m, Zusatzluftschieber m
ball valve   Kugelventil n
check valve   Sperrventil n, Rückschlagventil n
disk valve   Tellerventil n
distribution valve   Steuerventil n
distributor slide valve   Steuerschieber m
distributor valve   Verteilerventil n
drain valve   Entwässerungsventil n
exhaust valve   Auslaßventil n, Auspuffventil n
flat slide valve   Flachschieber m
flexible tube valve   Schlauchventil n
float-needle valve   Schwimmernadelventil n
foot valve   Fußventil n
guide valve   Führungsventil n
hand brake valve   Handbremsventil n
inflation valve   Luftschlauchventil n
inlet valve   Einlaßventil n
intake valve   Einlaßventil n
magnet valve   Magnetventil n
mechanically operated valve   mechanisch betätigtes Ventil n
multiway valve   Mehrwegventil n
non-return valve   Rückströmventil n, Rückschlagventil n
outlet valve   Ablaßventil n
overflow valve   Überströmventil n
overhead valve   hängendes (obengesteuertes) Ventil n
overload valve   Überdruckventil n
partial-load valve   Teillastventil n
preheating valve   Vorwärmklappe f
pressure valve   Druckventil n
quick release valve   Schnell-Löseventil n
reducing valve   Drosselventil n
regulating valve   Regelventil n, Regulierventil n
release valve   Auslöseventil n
relief valve   Entlastungsventil n
safety valve   Sicherheitsventil n

starting valve   Startventil n
stop valve   Absperrventil n
suction valve   Saugventil n, Ansaugventil n
three-way valve   Dreiwegventil n
throttle valve   Drosselklappe f
trailer-brake valve   Anhängerbremsventil n
vane-type pump   Flügelzellenpumpe f
vanity mirror   Make-up-Spiegel m
vaporizer   Verdampfer m
vapo[u]r lock   Blasenbildung f *in (Kraftstoffleitungen)*
variable   variabel, veränderlich
variable-speed governor   Verstellregler m
variable-stroke engine   Motor m mit veränderlichem Hub
varnish   Lack m
vehicle   Fahrzeug n
vehicle body   Karosserie f, Fahrzeugaufbau m, Fahrzeugkarosserie f
vehicle clearance circle   Wendekreis m
vehicle dynamics   Fahrzeugdynamik f
vehicle equipment   Fahrzeugausrüstung f
vehicle handling   Fahrverhalten n
vehicle identification number   Fahrgestellnummer f
vehicle registration   Fahrzeugzulassung f, Fahrzeugschein m,
Schein m, Fahrzeug-Zulassungspapiere npl
vehicle safety   Fahrzeugsicherheit f
motor vehicle   Kraftfahrzeug n
vent   Luftdüse f, Luftklappe f, Ausstellfenster n
vent cap   Entlüftungskappe f
vent line (pipe)   Entlüftungsleitung f, Entlüftungsgrohr n
vent screw   Entlüftungsschraube f
vent valve   Entlüftungsventil n
vent window   Ausstellfenster n, Drehfenster n, Schwenkfenster n,
Klappfenster n
air vent   Entgasungsöffnung f, Luftloch n
air vent hose   Entlüfterschlauch m
air vent screw   Entlüftungsschraube f *(Batterie)*
air vent valve   Entlüftungsventil n
ventilation   Lüftung f, Durchlüftung f, Entlüftung f
ventilation fan   Belüftungsgebläse n, Luftgebläse n
ventilation flap   Belüftungsklappe f, Luftklappe f

ventilation hole   Entlüftungsbohrung f; Belüftungsbohrung f
ventilator   Lüfter m *(s.a. fan)*
  ventilator hood   Lüfterhaube f
  ventilator window   Ausstellfenster n
  radial ventilator   Radiallüfter m
venturi tube   Venturirohr n, Zerstäuber m, Mischrohr n, Vergaserlufttrichter m
vertical   vertikal
  vertical seat adjustment   vertikale Sitzverstellung f, Sitzhöhenverstellung f
  vertical shaft   Königswelle f
  vertical valve   stehendes Ventil n
vessel   Behälter m
vibration   Schwingung f
  vibration absorber (damper)   Schwingungsdämpfer m
view   Blick-
  rear-view mirror   Rückspiegel m
visibility   Sicht f
vision   Sicht f
  clear vision screen   Klarsichtscheibe f
visor   Blendschutz m
  sun visor   Sonnenblende f
voltage   Spannung f
  voltage control   Spannungsregelung f
  voltage drop   Spannungsabfall m
  voltage increase   Spannungserhöhung f
  voltage regulator   Spannungsregler m, Regler m
  alternating voltage   Wechselspannung f
  battery voltage   Batteriespannung f
  charging voltage   Ladespannung f
  discharge voltage   Entladespannung f
  ignition voltage   Zündspannung f
  loss of voltage   Spannungsverlust m
  nominal voltage   Nennspannung f
  rated voltage   Nennspannung f
  spark-over voltage   Überschlagspannung f
  supply voltage   Anschlußspannung f
voltmeter   Spannungsmesser m
volume   1. Rauminhalt m, Volumen n; 2. Lautstärke f *(Radio)*

sound volume   Lautstärke f
swept volume   Hubraum m, Hubvolumen n
wall   Wand f
   front wall   Vorderwand f
   rear wall   Rückwand f, Hinterwand f
Wankel engine   Wankelmotor m
   Wankel rotary combustion engine   Wankelmotor m
   twin-rotor Wankel engine   Doppelläufer-Wankelmotor m
warm-up/to   warmlaufen
warm   warm, Warm-
   warm air duct   Warmluftkanal m
   warm air flap   Warmluftklappe f
   warm air hose   Warmluftschlauch m
   warm air regulator   Warmluftregler m
   warm-up control   Warmlaufregelung f, Anwärmregelung f
   warm-up control device   Warmlaufregler m, Anwärmregler m,
   Warmlaufeinrichtung f, Anwärmeinrichtung f
   warm-up device   Warmlaufeinrichtung f
   warm-up phase   Warmlaufphase f, Anwärmphase f
warn/to   warnen
warning   Warn-
   warning board   Warntafel f
   warning device   Warnanlage f; Warndreieck n
   warning lamp   Kontrollampe f, Warnleuchte f
   warning light   Kontrolleuchte f, Warnleuchte f
   warning triangle   Warndreieck n
   flasher warning light   Blinkkontrollampe f
   hand brake warning lamp   Handbremse-Kontrollampe f
   ignition warning light   Zündkontrollampe f
   main beam warning lamp   Fernlichtkontrollampe f
   petrol warning light   Benzinkontrollampe f, Benzinkontrolleuchte f
wash/to   waschen
wash-wiper   Scheibenwisch- und -waschanlage f
washer   1. Beilagscheibe f, Unterlegscheibe f; 2. Scheibenwaschanlage f
   washer pump   Scheibenwascherpumpe f, Wascherpumpe f
   washer reservoir   Wasserbehälter m für Scheibenwaschanlage
   lock washer   Sicherungsscheibe f
   spring washer   Federring m, Federscheibe f
   thrust washer   Anlaufscheibe f, Druckscheibe f

toothed washer   Zahnscheibe f
windshield washer   Scheibenwaschanlage f
washing system   Scheibenwaschanlage f
water   Wasser n
　water circulation   Wasserumlauf m
　water circulation cooling   Wasserumlaufkühlung f
　water-cooled engine   wassergekühlter Motor m
　water cooling   Wasserkühlung f
　water-drain channel   Ablaufblech n
　water drain hole   Wasserablaufloch n *(Karosserie)*
　water drain plug   Wasserablaßschraube f,
　Kühlwasserablaßschraube f, Kühlmittelablaßschraube f
　water drain tap   Wasserablaßhahn m, Kühlwasserablaßhahn m,
　Kühlmittelablaßhahn m
　water heated induction manifold   wasserbeheizter
　Einlaßkrümmer m, wasserbeheizter Ansaugkrümmer m,
　wasserbeheiztes Saugrohr n
　water inlet connection   Kühlwassereinlaufstutzen m
　water inlet pipe   Wassereinlaufstutzen m
　water inlet tube   Wassereinlaufrohr n
　water jacket   Wassermantel m, Kühlwassermantel m
　water level   Wasserstand m, Kühlwasserstand m
　Kühlflüssigkeitsstand m, Kühlmittelstand m
　water pipe   Wasserleitung f, Kühlwasserleitung f
　water pump   Wasserpumpe f
　water pump housing   Wasserpumpengehäuse n
　water pump shaft   Wasserpumpenwelle f
waterproof   wasserdicht, wasserbeständig
water seal   Dichtleiste f, Dichtstreifen m, Regenleiste f, Dichtring m,
　Abdichtgummi m
way   Weg m
　three-way valve   Dreiwegventil n
wear   Abnutzung f, Verschleiß m
　wear indicator   Abnutzungsanzeiger m, Verschleißanzeiger m
weather strip   Fensterdichtprofil n, Dichtstreifen m
　door weather strip   Türdichtungsstreifen m
webbing   Gurtband n, Gewebeband n
wedge   Keil m, Unterlegkeil m
　wedge shape   Keilform f

wedge-shaped combustion chamber keilförmiger Brennraum m
slot wedge Nutverschlußteil m
weight Gewicht n
  weight reduction Gewichtsreduzierung f
  weight saving Gewichtseinsparung f
  centrifugal weight Fliehmasse f
  dead weight Eigenmasse f, Leermasse f
weld/to schweißen
welding Schweißen n
  welding bolt Einschweißbolzen m, Schweißbolzen m
  welding joint Schweißnaht f
  welding seam Schweißnaht f
well Felgenbett n, Tiefbett n
  well base Tiefbett n
  well-base rim Tiefbettfelge f
  well radius Schulterradius m, Tiefbettradius m
wet naß
  wet air-cleaner Naßluftfilter n,m
  wet air-filter Naßluftfilter n,m
  wet battery Akkumulatorbatterie f
  wet filter Naßfilter n,m
  wet liner nasse Laufbuchse f *(unmittelbar vom Kühlwasser umspült)*
  wet sump Ölsumpf m
wheel Rad n, Laufrad n
  wheel alignment Radeinstellung f, Radvermessung f, Radfluchtung f, Spureinstellung f
  wheel angle Radeinschlag m, Lenkeinschlag m, Lenkwinkel m
  wheel arch Radkasten m
  wheel balancing Radauswuchten n, Radauswuchtung f
  wheel base Radstand m, Achsabstand m
  wheel bearing Radlager n
  wheel bolt Radbolzen m
  wheel brake cylinder Radbremszylinder m, Radzylinder m
  wheel camber Radsturz m, Sturz m
  wheel cap Radkappe f, Radzierblende f, Zierblende f, Zierkappe f
  wheel caster Nachlauf m
  wheel center Radmittelpunkt m, Radmitte f, Radscheibe f, Schüssel f

wheel chock   Unterlegkeil m, Radkeil m
wheel cover   Radkappe f, Radzierblende f, Zierblende f, Zierkappe f
wheel cylinder   Radzylinder m, Radbremszylinder m
wheel deflection   Radausschlag m
wheel disk   Radscheibe f, Radschüssel f, Schüssel f
wheel house   Radhaus n
wheel hub   Radnabe f
wheel hub cap (cover)   Radnabendeckel m
wheel imbalance   Radunwucht f
wheel lockup   Blockieren n eines Rades
wheel nut   Radmutter f
wheel offset   Lenkrollradius m
wheel opening   Radausschnitt m
wheel rake   Radsturz m, Sturz m
wheel recess   Radmulde f
wheel rim   Radfelge f
wheel shimmy   Radflattern n
wheel slip   Radschlupf m, Reifenschlupf m, Schlupf m, Umfangsschlupf m, Schlupf m in Umfangsrichtung, Radschlupf m in Umfangsrichtung
wheel slip brake control   Bremsschlupfregelung f
wheel slip control   Radschlupfregelung f, Bremsschlupfregelung f
wheel spider   Radstern m, Speichenstern m
wheel suspension   Radaufhängung f
wheel toe   Radspurwinkel m, Spurwinkel m, Radspur f, Spur f
wheel track   Spurweite f, Spur f, Radspur f
all-wheel drive   Allradantrieb m
chain tension wheel   Kettenspannrad n
chain wheel   Kettenrad n
crown wheel   großes Differentialantriebskegelrad n, Tellerrad n
disk wheel   Scheibenrad n
driving wheel   Antriebsrad n
fan wheel   Lüfterrad n
fifth-wheel load   Aufsattellast f, Aufsatteldruck m
fifth-wheel steering   Drehschemellenkung f
four-wheel drive   Vierradantrieb m, Geländeantrieb m
front wheel drive   Vorderradantrieb m
gear wheel   Zahnrad n, Getrieberad n
long wheel base   großer Radabstand m

planetary wheel   Planetenrad n
rear wheel drive   Hinterradantrieb m
spur wheel   Stirnrad n
steering wheel   Lenkrad n
tightening wheel   Spannrad n
toothed wheel   Zahnrad n
trailing wheel   Laufrad n
worm-and-wheel steering gear   Schneckenlenkung f
wheeling   Lauf-
free-wheeling   Freilauf m
free-wheeling bush   Freilaufbuchse f
wide   breit
wide-base rim   Breitfelge f
wide-base tire   Breitreifen m, breiter Reifen m
wide beam headlamp   Breitstrahler m
wide-open throttle kickdown valve   Übergasventil n,
Kick-down-Ventil n
wide range headlamp   Breitstrahler m
wide tire   Breitreifen m, breiter Reifen m
width   Breite f
track width   Spurbreite f, Spurweite f
wind/to   aufspulen, wickeln
winding   Wicklung f
window winding mechanism   Fensterheber m
window   Fenster n, Scheibe f
window blind   Fensterjalousie f
window channel   Fensterschacht m, Fensterführung f,
Fensterscheibenführung f
window control   1. Fensterkurbel f; 2. Schalter m für
automatischen Fensterheber
window crank   Fensterkurbel f
window curvature   Fensterkrümmung f
window guide   Fensterführung f
window lift   Fensterkurbel f
window lifter   Fensterheber m
window lifter rail   Fensterheberschiene f
window pane   Fensterscheibe f, Scheibe f
window pillar   Fensterpfosten m, Fenstersäule f, Fensterholm m
window post   Fensterpfosten m, Fenstersäule f, Fensterholm m

window seal Fensterdichtung f, Fensterabdichtung f, Fenstereinfassung f

window winding mechanism Fensterheber m

crank-operated window Kurbelfenster n

ventilator window Ausstellfenster n

windscreen Windschutzscheibe f

windscreen frame Windschutzrahmen m

windscreen misting Beschlagen n der Windschutzscheibe

windscreen washer Scheibenwascher m

windscreen weather seal (weather strip)
Windschutzscheibeneinfassung f, Windschutzscheiben-
abdichtung f, Windschutzscheibendichtung f

windscreen wiper Scheibenwischer m,
Windschutzscheibenwischer m

windscreen wiper equipment Scheibenwischeranlage f

windscreen wiper linkage Scheibenwischergestänge n

windscreen wiper motor Scheibenwischermotor m

windshield Windschutzscheibe f

windshield washer Scheibenwaschanlage f

windshield weather seal (weather strip)
Windschutzscheibeneinfassung f, Windschutzscheiben-
abdichtung f, Windschutzscheibendichtung f

windshield wiper Scheibenwischer m

wind tunnel Windkanal m

wing Kotflügel m, Flügel m

wing nut Flügelmutter f

front wing Vorderkotflügel m, vorderer Kotflügel m

window wing Fensterflügel m

wipe/to wischen

wash-wiper Scheibenwisch- und -waschanlage f

wiped area Wischfeld n

wiper Scheibenwischer m, Wischer m

wiper arm Wischerarm m

wiper bar Wischerstange f

wiper blade Scheibenwischerblatt n, Wischerblatt n

wiper linkage Wischergestänge n

wiper motor Scheibenwischermotor m

wiper pattern Wischerfeld n

wiper rubber Wischergummi m

wiper switch Wischerschalter m
wiper system Wischeranlage f, Scheibenwischanlage f
　rotary wiper motor Umlaufwischermotor m
　windscreen (windshield) wiper Scheibenwischer m
wiping speed Wischgeschwindigkeit f
wire Draht m, Leiter m, Leitung f
　wire bow Drahtbügel m
　wire circlip Drahtsprengring m
　earth (ground) wire Masseleitung f
　hot wire Hitzdraht m
　spark-plug wire Zündkabel n
wiring Schaltung f, Verkabelung f
　wiring diagram Schaltplan m
　wiring harness Kabelsatz m, Kabelbaum m
　wiring scheme Schaltplan m
wishbone Querlenker m, Trapezlenker m, Dreieckslenker m
wobble/to flattern *(Rad)*
woodruff key Woodruff-Keil m
work/to arbeiten
working Arbeits-
　working capacity Arbeitsvermögen n
　working chamber Arbeitskammer f *(Kreiskolbenmotor)*
　working cylinder Arbeitszylinder m *(Servolenkung)*
　working pressure Arbeitsdruck m, Betriebsdruck m
　working process Arbeitsgang m
workshop Werkstatt f
worm Schnecke f
　worm-and-roller steering Schneckenlenkung f
　worm-and-sector steering gear Schneckenlenkung f
　worm-and-wheel steering Schneckenlenkung f
　worm gear Schneckengetriebe n, Schraubenrad n
　worm gear steering Schneckenlenkung f
　steering worm Lenkschnecke f
wrap-around windshield Rundsichtverglasung f,
　Vollsichtverglasung f
wrecker Abschleppwagen m
yaw/to gieren *(Schlingerbewegung)*
　yaw moment Giermoment n
yoke Joch n, Polgehäuse n, Gabel f, Gabelkopf m

yoke end   Gabelkopf m
universal-joint yoke   Kreuzgelenkgabel f
zinc   Zink n
zinc coat   Zinkbeschichtung f
zinc-coated   verzinkt